ELECTRICAL THEORY

By Tom Henry

Copyright©2002 by Tom Henry. All rights reserved. No part of this publication may be reproduced in any form or by any means: Electronic, mechanical, photocopying, audio or video recording, scanning, or otherwise without prior written permission of the copyright holder.

While every precaution has been taken in the preparation of this course to ensure accuracy and reliability of the information, instructions, and directions, it is in no way to be construed as a guarantee. The author and publisher assumes no responsibility for errors or omissions. Neither is any liability assumed from the use of the information contained herein in case of misinterpretations, human error, or typographical errors.

National Electrical Code® and NEC® are Registered Trademarks of the National Fire Protection Association, Inc., Quincy, MA.

14th Printing July 2011   **ISBN   978-0-945495-56-7**

 ENRY PUBLICATIONS SINCE 1985

CONTENTS

ELECTRICITY

It is not a solid, liquid, nor gas; something having no weight, being invisible, occupying no space, moving at enormous speed, but yet a normal part of nature.

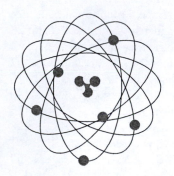

To learn electricity, the invisible force that became the major means of making man the master of the earth, you must first learn the methods of producing and controlling it and learn the rules or laws that apply to the *behavior* of electricity.

All the effects take place in a tiny particle called the electron. Since you can't see an electron, this book will be very helpful as it will show, in motion, its behavior, called electron theory.

Electricity is produced by very tiny particles called *electrons* and *protons*.

These electrons and protons are too small to see, but they exist in all *matter*.

MATTER

HAS WEIGHT

TAKES UP SPACE

SOLID
LIQUID
GAS

Matter is defined as anything that has *weight* and takes up *space*. It can be in the form of a solid, liquid, or gas.

First we must know something about the atomic structure of matter.

Atoms are the basic materials that make up all matter. Everything we see is made from atoms. There are over 100 atoms such as oxygen, hydrogen, nitrogen, aluminum, copper, silver, gold, mercury, lead, sodium, and chlorine to name a few.

There are many more materials than atoms. Atoms can be combined to produce materials. An example is water which is a compound made from hydrogen and oxygen. Although hydrogen and oxygen are both gases that when combined (H^2O) can produce water; a liquid. Through the activity of the electrons, atoms combine to form a *molecule* of material.

The molecule is the smallest particle that a compound can be reduced to before it breaks down into its atoms.

MOLECULE OF SALT = **ATOM CHLORINE** + **ATOM SODIUM**

An example is salt which is produced by the two atoms chlorine and sodium. If a grain of salt was broken in half and that half was broken in half and you continued to break it in half until it could still be recognized as salt we would have a *molecule* of salt. If we broke it in half again we would have reduced it back to its two atoms of chlorine and sodium.

The smallest particle that an element can be reduced to and still keep the properties of that element is called an *atom*.

Atom is the Greek word for indivisible (not able to be divided).

An atom contains three types of subatomic particles: Electrons, protons, and neutrons. The protons and neutrons are located in the center of the atom (nucleus), and the electrons travel around them in orbit.

ATOM

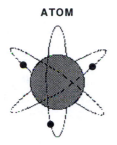

Atoms resemble a miniature solar system. The electron revolves around the nucleus in a fixed orbit just as planets revolve around the sun.

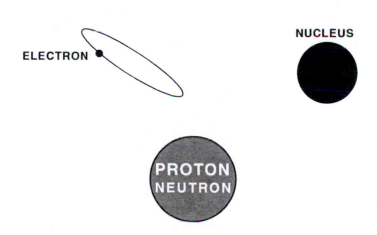

The nucleus, which contains the protons and neutrons, is so small that if it were enlarged to the size of a pinhead and the pinhead were enlarged in comparison to the enlargement of the atom, the pinhead would have a diameter of 92,000,000 miles. Approximately the distance between the earth and the sun.

The neutron is a particle and has about the same mass as the proton, but has no charge at all.

It is considered as an electron and proton combined, since it has no charge it is considered *neutral*. Neutrons are not too important to the electrical nature of atoms.

NEUTRONS = 0
PROTONS = +

Since neutrons are neutral and protons are *positive*, the nucleus of the atom is *positive*.

Even though the proton is only 1/3 the diameter of an electron it is 1840 times heavier. It is very difficult to dislodge a proton from the nucleus of an atom.

The nucleus is like a package made of smaller parts. About 30 different kinds of particles might be present, but only two kinds are important; the proton and the neutron.

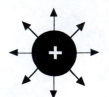

In electrical theory, protons are considered permanent in the nucleus. Protons do not participate in the flow of electrical energy. The proton has a positive charge.

The positive lines of force of a proton go straight out in all directions.

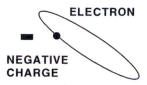

ELECTRON

NEGATIVE CHARGE

Electrons orbit around the nucleus of the atom and have negative (-) charges. Electrons are easy to move since they are lighter in weight. They are the particles that participate in the flow of electrical energy.

To describe an electron in size, if it was the size of a ping-pong ball, the ping-pong ball in comparison would be the size of earth's orbit; 186,000,000 miles in diameter.

The electron is capable of exerting a tremendous force; perhaps the greatest force known to man. Remember the atomic bomb's explosive power is due to the sudden release of atomic energy resulting from splitting the nuclei of uranium.

It has been estimated that if two grams (0.070 ounce) of electrons could be collected into two equal spheres and these spheres were a distance of 0.39" apart, they would repel each other with a force of 320,000,000,000,000,000,000,000 tons. A force greater than the weight of the water in all the oceans of the world.

The number of protons in the nucleus determines how one atom differs from another atom.

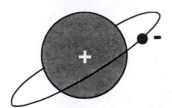

Hydrogen has one proton whereas carbon has six protons. The number of electrons will be balanced.

ALUMINUM = 13 PROTONS

COPPER = 29 PROTONS

Normally there is one proton for each electron in the entire atom which balances, thus the atom is electrically neutral.

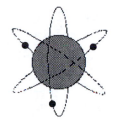

The electrons do not fall into the nucleus even though they are strongly attracted to it. Their motion prevents it because of their centrifugal force. The protons do not move because they are extremely heavy.

The orbits of these electrons are arranged in "*shells*" about the nucleus, each shell having a *definite* maximum capacity of electrons. Some atoms have up to seven shells.

Hydrogen has one orbital shell, carbon has two orbital shells, aluminum has three orbital shells whereas copper has four orbital shells.

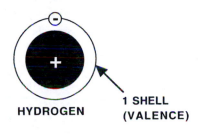

HYDROGEN

1 SHELL (VALENCE)

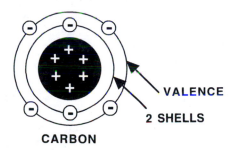

VALENCE

2 SHELLS

CARBON

The capacity of successive shells from the nucleus out is 2, 8, 18, and 32 electrons; however the *outermost shell* (the valence shell) never contains more than 8 electrons.

It is the outermost shell which determines the chemical *valence* of an atom and its principal characteristics.

The outermost shell is the most important for electricity, since it is the only one from which electrons are relatively *easily* dislodged to become "*free*" electrons capable of carrying current in a conductor.

For the study of electricity, we are concerned mainly with the one electron in orbit on the outside shell of the copper atom.

Copper is the conductor mainly used to carry electrical current, an aluminum conductor is the next most widely used.

Electrons that orbit close to the nucleus cannot be forced out easily from their orbits, and are considered "*bound*" to the atom.

ELECTRON SHELLS

Atomic No.	Element	Electrons per shell				
		1	2	3	4	5
1	Hydrogen, H	1				
2	Helium, He	2				
3	Lithium, Li	2	1			
4	Beryllium, Be	2	2			
5	Boron, B	2	3			
6	Carbon, C	2	4			
7	Nitrogen, N	2	5			
8	Oxygen, O	2	6			
9	Fluorine, F	2	7			
10	Neon, Ne	2	8			
11	Sodium, Na	2	8	1		
12	Magnesium, Mg	2	8	2		
13	Aluminum, Al	2	8	3		
14	Silicon, Si	2	8	4		
15	Phosphorus, P	2	8	5		
16	Sulfur, S	2	8	6		
17	Chlorine, Cl	2	8	7		
18	Argon, A	2	8	8		
19	Potassium, K	2	8	8	1	
20	Calcium, Ca	2	8	8	2	
21	Scandium, Sc	2	8	9	2	
22	Titanium, Ti	2	8	10	2	
23	Vanadium, V	2	8	11	2	
24	Chromium, Cr	2	8	13	1	
25	Manganese, Mn	2	8	13	2	
26	Iron, Fe	2	8	14	2	
27	Cobalt, Co	2	8	15	2	
28	Nickel, Ni	2	8	16	2	
29	Copper, Cu	2	8	18	1	
30	Zinc, Zn	2	8	18	2	
31	Gallium, Ga	2	8	18	3	
32	Germanium, Ge	2	8	18	4	
33	Arsenic, As	2	8	18	5	
34	Selenium, Se	2	8	18	6	
35	Bromine, Br	2	8	18	7	
36	Krypton, Kr	2	8	18	8	
37	Rubidium, Rb	2	8	18	8	1
38	Strontium, Sr	2	8	18	8	2
39	Yttrium, Y	2	8	18	9	2
40	Zirconium, Zr	2	8	18	10	2
41	Niobium, Nb	2	8	18	12	1
42	Molybdenum, Mo	2	8	18	13	1
43	Technetium, Tc	2	8	18	14	1
44	Ruthenium, Ru	2	8	18	15	1
45	Rhodium, Rh	2	8	18	16	1
46	Palladium, Pd	2	8	18	18	0
47	Silver, Ag	2	8	18	18	1
48	Cadmium, Cd	2	8	18	18	2
49	Indium, In	2	8	18	18	3
50	Tin, Sn	2	8	18	18	4
51	Antimony, Sb	2	8	18	18	5
52	Tellurium, Te	2	8	18	18	6

Atomic No.	Element	Electrons per shell						
		1	2	3	4	5	6	7
53	Iodine, I	2	8	18	18	7		
54	Xenon, Xe	2	8	18	18	8		
55	Cesium, Cs	2	8	18	18	8	1	
56	Barium, Ba	2	8	18	18	8	2	
57	Lanthanum, La	2	8	18	18	9	2	
58	Cerium, Ce	2	8	18	19	9	2	
59	Praseodymium, Pr	2	8	19	20	9	2	
60	Neodymium, Nd	2	8	19	21	9	2	
61	Promethium, Pm	2	8	18	22	9	2	
62	Samarium, Sm	2	8	18	23	9	2	
63	Europium, Eu	2	8	18	24	9	2	
64	Gadolinium, Gd	2	8	18	25	9	2	
65	Terbium, Tb	2	8	18	26	9	2	
66	Dysprosium, Dy	2	8	18	27	9	2	
67	Holmium, Ho	2	8	18	28	9	2	
68	Erbium, Er	2	8	18	29	9	2	
69	Thulium, Tm	2	8	18	30	9	2	
70	Ytterbium, Yb	2	8	18	31	9	2	
71	Lutetium, Lu	2	8	18	32	9	2	
72	Hafnium, Hf	2	8	18	32	10	2	
73	Tantalum, Ta	2	8	18	32	11	2	
74	Tungsten, W	2	8	18	32	12	2	
75	Rhenium, Re	2	8	18	32	13	2	
76	Osmium, Os	2	8	18	32	14	2	
77	Iridium, Ir	2	8	18	32	15	2	
78	Platinum, Pt	2	8	18	32	16	2	
79	Gold, Au	2	8	18	32	18	1	
80	Mercury, Hg	2	8	18	32	18	2	
81	Thallium, Tl	2	8	18	32	18	3	
82	Lead, Pb	2	8	18	32	18	4	
83	Bismuth, Bi	2	8	18	32	18	5	
84	Polonium, Po	2	8	18	32	18	6	
85	Astatine, At	2	8	18	32	18	7	
86	Radon, Rn	2	8	18	32	18	8	
87	Francium, Fr	2	8	18	32	18	8	1
88	Radium, Ra	2	8	18	32	18	8	2
89	Actinium, Ac	2	8	18	32	18	9	2
90	Thorium, Th	2	8	18	32	19	9	2
91	Protactinium, Pa	2	8	18	32	20	9	2
92	Uranium, U	2	8	18	32	21	9	2
93	Neptunium, Np	2	8	18	32	22	9	2
94	Plutonium, Pu	2	8	18	32	23	9	2
95	Americium, Am	2	8	18	32	24	9	2
96	Curium, Cm	2	8	18	32	25	9	2
97	Berkelium, Bk	2	8	18	32	26	9	2
98	Californium, Cf	2	8	18	32	27	9	2
99	Einsteinium, E	2	8	18	32	28	9	2
100	Fermium, Fm	2	8	18	32	29	9	2
101	Mendelevium, Mv	2	8	18	32	30	9	2
102	Nobelium, No	2	8	18	32	31	9	2
103	Lawrencium, Lw	2	8	18	32	32	9	2

The Electron Shell Chart shows *copper* with 29 electrons and four shells. The outer valence shell has only one electron.

Hydrogen is the lightest atom in weight. It has only one shell with one electron. Aluminum is ranked #13 in weight as it has 13 electrons and three shells. The outer valence shell has three electrons.

Uranium is ranked #92 in weight and is considered heavy in weight compared to the other atoms.

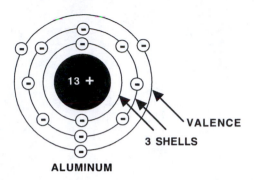

ALUMINUM

Aluminum has 13 positive protons in the nucleus and 13 negative electrons orbiting in 3 shells around the nucleus. The shell orbiting closest to the nucleus has 2 electrons. The middle orbiting shell has 8 electrons and the *outermost* shell (the valence) has 3 orbiting electrons.

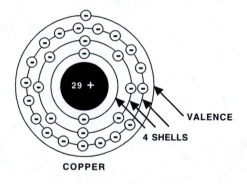

COPPER

Copper has 29 positive protons in the nucleus and 29 negative electrons orbiting in 4 shells around the nucleus. The shell orbiting closest to the nucleus has 2 electrons. The next shell outward has 8 orbiting electrons. The third outward shell has 18 electrons and the outermost shell (the valence) has 1 orbiting electron.

When an orbital electron is removed from an atom it is called a *free electron*. These dislodged electrons can exist by themselves outside of the atom, and it is these free electrons which are responsible for *electricity*.

Some of the electrons of certain metallic atoms such as copper or aluminum are so loosely bound they are comparatively free to move to another atom when a very small force or amount of energy is used to remove these free electrons. It is these free electrons that constitute the flow of an electric current in a conductor.

The force used to free an electron can be by friction, magnetism, chemical, heat, pressure, or light.

When force is applied to the valence electron(s), the energy is distributed evenly among all the electrons. The more valence electrons there are, the less energy each electron will receive. It is easier to free *one* valence electron than two valence electrons.

Conductors that carry current are made of materials that have *one*, *two*, or *three* valence electrons.

Elements that have *one* valence electron such as copper, silver, and gold make the best electrical conductors.

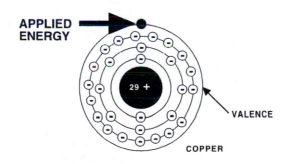

11

Aluminum with three valence electrons is not as good a conductor as copper which has only one valence electron.

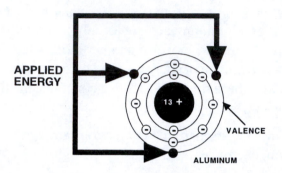

Insulators are elements that have *six or more* valence electrons. These electrons are very difficult to free and tend to catch any free electrons. An atom is completely stable when its outer valence is filled with 8 electrons and will *resist* any activity. The atom with 7 valence electrons is the most active, always looking to catch a free electron and is the *best* insulator. There is no such thing as a perfect insulator. Considered to be the best insulators are glass, mica, plastics, rubber, dry air, ceramics, and slate.

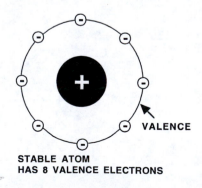

STABLE ATOM
HAS 8 VALENCE ELECTRONS

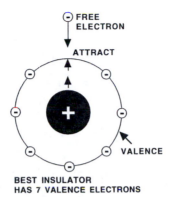

BEST INSULATOR
HAS 7 VALENCE ELECTRONS

Atoms that have 8 valence electrons are stable. Those that are less than half filled (the conductors) *release* electrons to an unstable valence shell. But those that are more than half filled (the insulators) are looking to *catch* more electrons to fill up the valence shell.

A semiconductor is a material that has some characteristics of both conductors and insulators. Semiconductors have become extremely important as the basis for transistors, diodes, and other solid-state devices used in electronics. They have four valence electrons.

An *ion* is an atom (or molecule) that has become electrically unbalanced by the loss or gain of one or more electrons. An atom that has *lost* an electron is called a *positive* ion, while an atom that has *gained* an electron is called a *negative* ion. When an atom loses an electron, its remaining orbital electrons no longer balance the positive charge of the nucleus, and the atom acquires a charge of +1. Similarly, when an atom gains an electron in some way, it acquires an excess negative charge of -1. The process of producing ions is called *ionization*.

An ion may be defined as a small particle of matter having a positive or negative charge.

Ionization does not change the chemical properties of an atom, but it does produce an electrical change.

The law of electrical charges is that a negative charge will *repel* another negative charge and a positive charge will *repel* another positive charge.

Like charges repel and unlike charges attract.

A negative charge will be *attracted* to a positive charge.

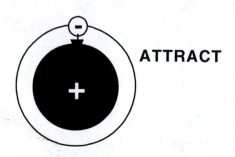

A horseshoe magnet can not be pushed together if the ends are likes as shown below on the left.

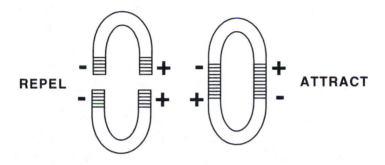

REPEL   ATTRACT

Energy excites or agitates the atoms causing electrons in the outer orbit to break away.

Friction is a form of energy that can free the electron in the atom. Rubbing two materials together can produce this excess or lack of electrons to cause static electricity.

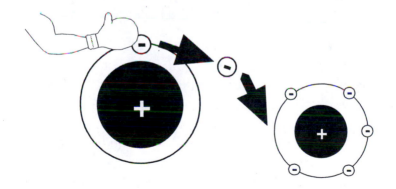

Some materials which easily build up static electricity are nylon, rayon, silk, flannel, hard rubber, and glass.

When glass is rubbed with silk, the glass rod loses electrons and becomes positively charged and the silk negatively charged. When hard rubber is rubbed with fur, the fur loses electrons to the rubber rod, the rod becomes negatively charged and the fur positively charged.

The original source of static electricity is friction, but you will find in some cases that a static charge may transfer from one material to another without much friction.

Static electricity has limited practical use, its presence is generally unpleasant and sometimes even dangerous. You probably have experienced on a dry day walking across nylon carpet and touching a metal switch cover or door knob and getting an unpleasant shock.

When any two materials are rubbed together electrons are discharged by the friction. If you experiment with a few different combinations of materials you will find that some work well and some don't. If good conductors are used, it is difficult to notice any friction. The reason is that equalizing charges will flow easily in and between the conductor materials. They equalize almost as fast as they are created. A static charge is easier to obtain by rubbing a hard nonconducting material against a soft nonconductor. Electrons are rubbed off one material and onto the other material.

When the glass rod is rubbed with the silk, the silk accumulates electrons. Since both the glass and the silk are poor conductors, little equalizing current can flow, and an electrostatic charge is built up. When the charge is great enough, equalizing currents will flow regardless of the material's poor conductivity. These currents will cause visible sparks in the darkness and produce a cracking sound.

Lightning is an example of the discharge of static electricity generated from the friction between a cloud and the surrounding air. The lightning is an arc type electrical discharge that neutralizes the positive and negative charges that have built up.

Lightning rods are installed at the highest point of a building to attract the electrons and send them to planet earth for discharge.

So far we have been learning about electricity and how an electric charge is produced. This is called *static electricity,* which is an electrical charge at *rest*. For electrical energy to do work, the electricity must be put in *motion*. This is called electric *current*. Electric current is developed when many free electrons in a conductor are moved in the *same direction.*

A free electron moves in a random manner throughout the conductor until it is *attracted* to an atom which has lost an electron.

CONDUCTOR

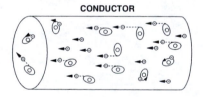

When free electrons move randomly in all directions, that undirected action does not produce any useful results.

When we can channel all the electrons to move together in *one direction* we then have a flow of current or usable electricity.

One way of getting the electrons to go in one direction is by connecting a conductor to an electrical energy source.

To make current flow, a potential difference must be maintained. When current flows, energy is required to maintain this potential difference and work must be done.

Current flow is when free electrons are guided from the energy source in an orderly fashion, atom to atom through the conductor back to the energy source.

Although the electrons drift through the conductor at a relatively low speed, the impulse is transmitted at the speed of light. Current will continue to flow as long as there is a *potential difference*. The conductor itself remains electrically neutral, since electrons are neither gained nor lost by the atoms within the conductor. Electrons enter the conductor from the negative side of the energy source and an equal number of electrons are given up by the other end of the conductor to the positive source of energy. The free electrons present within the conductor act simply as current carriers, which are continually being replaced, but none are lost.

The force that causes free electrons to move in a conductor as an electric current may be called a difference in potential, voltage, or electromotive force (emf).

When a difference in potential exists between two charged bodies that are connected by a conductor, electrons will flow along the conductor. This flow is from the negatively charged body to the positively charged body until the two charges are equalized and the potential difference no longer exists.

The two water tanks below are connected by a pipe and a valve. The first tank is full of water, but with the valve open water flows into the second tank until the water in both tanks are the same level. Water stops flowing because there is *no longer a difference in water pressure* between the two tanks.

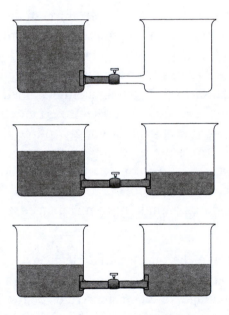

Remember how small an atom is, the approximate number of atoms in a cubic centimeter (0.061 cubic inch) of copper is 1,000,000,000,000,000,000,000,000. So you can see there is an enormous number of free electrons moving about at random through the conductor.

There are several sources of energy to release the free electrons from their orbit as shown below.

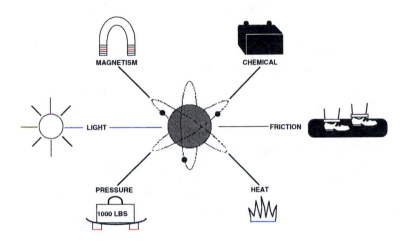

The battery uses chemical action to build up an excess of electrons on its negative terminal. Its positive terminal has a build up of atoms with missing electrons. Connecting a conductor to the battery will cause most of the free electrons to be attracted by the positive terminal of the battery. The electrons move in one direction from *negative to positive*. Electrons continue to flow as long as the battery has a chemical unbalance and there is a closed circuit.

In electron theory current flow is always from a negative (-) charge to a positive (+) charge. Electrical *equipment* uses the concept of *conventional current flow* which is the current flow is from positive to negative.

The cell consists of two electrodes which are partially submerged in a liquid solution that is referred to as the *electrolyte*. The electrolyte is a weak solution of acid, base, or salt in water.

The negative ions and the positive ions float in the electrolyte. Because of the chemical reaction, positive ions are attracted to one electrode and negative ions to the other. With one electrode negatively charged and the other electrode positively charged, a difference of potential (voltage) exists across the electrodes.

 If a conductor and a lamp are connected between the two electrodes, electrons will flow between the electrodes and the bulb will light. This is a *closed* circuit which allows current to flow from the source through the conductor to the load and back to the source.

Applied pressure to certain materials can cause an electrical charge. This is called *piezoelectricity*. A Greek word for pressure. This is used in some microphones and phonographs.

Heat can produce electricity. This is called *thermoelectricity*. The more heat that is applied to the junction of two metals the greater the charge that will be built up. This device is called a thermocouple used in instrumentation. Even when two dissimilar metals are joined together a transfer of electrons can take place.

Light is also a form of energy and is made up of small particles called *photons*. Photoemission, photovoltaic, and photoconduction are forms of *photoelectricity*.

In the previous pages we have discussed friction (static) and battery (chemical) as means for producing electron movement. The most widely used form of energy is *magnetism*. Magnetism will be covered as a separate subject later.

We must have a form of energy to make the electron move, and when it moves through a conductor electrical current will flow.

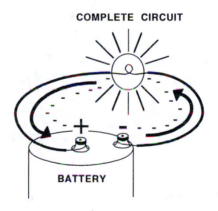

COMPLETE CIRCUIT

BATTERY

The circuit will have an energy source, conductors (which are insulated) to carry the current flow, a load (light bulb, etc.) and conductors to provide a path back to the source of energy.

Don't think of electrons coming into direct physical contact with one another. Remember, like charges *repel*, and all electrons are *negative*.

When a moving electron comes close to another electron, the second electron will be pushed away by the electric field of the first, without the two electrons themselves ever coming into contact.

Electrical current is the impulse that one electron transmits to another electron as it changes orbit. As the freed electron enters the new orbit its negative charge reacts with the negative charge of the electron already in that orbit. The second electron repeats the performance of the first as it encounters the next orbit. This process continues through the wire. The impulse of energy that is transferred from electron to electron is the electric current.

When current flow starts in a conductor, electrons start to move throughout the conductor at the *same time*, just as the cars of a long train start and stop at the same time.

The electrons move rapidly, but not very far. An electron that leaves the negative terminal of the battery might never get to the positive terminal. There are billions and billions of free electrons in the conductor. The entire group of electrons act together, like a single electron. While the impulse travels near the speed of light, the electrons just move and nudge each other along.

It has been estimated that a *single* electron m̠ ̠es rather slowly at a rate of approximately 3 inches per hour at one ampere of current flow. But, the *impulse* of electricity, is extremely fast. It is assumed that the speed of the electrical impulse is 186,000 miles per second, which is the speed of light. When one electron enters a conductor with billions of electrons, the impulse must be fast to knock one out the other end of the conductor and move billions to do so.

The difference of potential unit of measurement is called the volt. The unit to measure current flow is called the ampere. It is equal to 1 coulomb per second. We must measure the flow of current by measuring the number of electrons that flow past a given point in a given period of time. Since the coulomb is a measure of the number of electrons present, we can use it as a basis for the measurement of current flow. A coulomb is defined as about six and a quarter million, million, million electrons. An ampere is defined as 1 coulomb flowing in one second of time.

If you should attempt to count the electrons in one ampere of current for *one second*, you would have to count 1000 electrons per second for 190 million years without stopping.

It isn't necessary to remember the number of electrons per second in an ampere; however, it is important to remember that electrons in motion are current flow and that one ampere is the unit of measurement for current flow.

The Beginning

First there was lightning which lit up the sky. There was the attraction of rubbed amber for tiny bits of lint, dust, and feathers. But the human need to explain and justify was not, until our own times, satisfied by scientific reasoning.

For century after century man was knowledgeable of the barely noticeable forces; electrostatic attraction in amber and the magnetic attraction in lodestone, but went no further.

The earliest useful concept of electricity, among the early Greeks, was that it was not a thing but a property, an attribute of certain substances. Many knew of these amber and lodestone powers, but few asked the question, "why?" Not until the time of the early Greeks did man ask "why?"

Approximately seven centuries before Christ, Thales of Miletus, an astronomer offered a brief explanation: He suggested that these substances had a *soul* within them, that they were alive, because they could attract objects. This was the beginning of scientific curiosity and there was no better theory offered for another 2000 years.

Amber is not a stone, gem, or mineral. It is a harden tree resin that is millions of years old. Unlike other fossils, it is organic. It even burns.

Amber is often called "an organic jewel." Gold is the most predominant color, followed by yellow and red, but amber comes in dozens of colors, including rare greens and blues ranging from transparent to opaque.

Today amber exhibits display pieces that are 94 million years old and the largest piece of transparent amber ever found weighed 33 pounds.

The Greek word for amber is *elektron*. Rubbed amber would attract light objects.

The relationship between electricity and magnetism is the foundation to learning about electricity.

The story is told that in the year 500 B.C., a shepherd boy who lived in a Greek town named Magnesia often used a walking stick with an iron tip to help him climb the stony hills.

One day the tip of his walking stick actually stuck to a stone. He reached down and felt the stone. It didn't feel sticky, and nothing else was attracted to it. Only the iron tip of his stick clung to the rock. This rock is called a *lodestone*. Lodestone is an iron oxide, magnetite, grayish-black, naturally magnetized by earth itself.

The usual spelling "loadstone" suggests "the stone can lift a load."

In China about 376 B.C. Haung Ti, a Chinese general, employed the first practical application of magnetism (lodestone) in the compass. Travelers used it to find their way. Certainly Christopher Columbus made good use of it.

Earth's magnetic north and south poles are not located at the same place as the true, or geographic, north and south poles. The north pole on a compass is located in Canada about 1000 miles from the true north pole. The magnetic south pole is about 1,500 miles from the true south pole directly south of Australia.

Birds take advantage of Earth's magnetic field in flying south for the winter. Homing pigeons have a very small magnetic crystal located between the brain and the skull which provides them with a geomagnetic field for orientation.

Magnetism is all around you from the magnet holding the paper on the refrigerator, or the telephone ringing. While you're watching TV imagine a single dot racing back and forth across the front of your picture tube. This dot is directed and controlled by a magnetic field created by a deflection yoke on the back of your picture tube. Japan has a train that can float on a magnetic field at 300 miles per hour.

M agnetism is defined as an invisible force of attraction, like an *invisible glue*. Magnetism is like electricity; we cannot see it, but yet we can tell that it exists because it produces certain effects. There is no insulator for magnetism. It penetrates everything.

In the 13th century man no longer satisfied with past theories, sought objective knowledge supported by experiment to find clear observations on the properties of amber and lodestone.

In 1576 Jerome Cardan made clear in his discoveries that amber and lodestone attractions differed. This started the explosion of inquiry that in less than four centuries would carry our Western civilization to its present state of knowledge of electricity.

In 1600 an English physician, William Gilbert (1544-1603), was the first person to distinguish between electricity and magnetism. In his book he published his discovery that Earth itself is a magnet.

Gilbert found many substances that when rubbed, behaved like amber. He called them *electrics*. He is given credit for the word *electricity*.

The relationship between electricity and magnetism is the foundation to learning about electricity.

Magnetism affects all materials to a lesser or greater degree. This is called *permeability*. The dictionary defines this as to permeate is to saturate. Permeability is like the flow of water through a sand filled pipe. The capacity of a body of a porous sediment to transmit fluids depends on the closeness or connections of spaces between the pores along with the size of the area of the body. The ability of a substance to pass or conduct the lines of a magnetic force is called the permeability of a material.

Materials are classified into three groups; paramagnetic, diamagnetic, and *ferromagnetic*.
Paramagnetic and diamagnetic materials are those that become only slightly magnetized under a strong magnetic field as they have a very low permeability.

Ferromagnetic materials are the most important to electricity as they are relatively easy to magnetize such as iron, steel, and cobalt. Ferromagnetic materials all have a *high permeability.*

Because of the many uses of magnets, they are found in various shapes and sizes. The three primary classifications are bar magnets, horseshoe magnets, and ring magnets.

There are only two types of magnets: Permanent magnets and electromagnets. A magnet is normally made of iron.

Hard materials are used for permanent magnets such as the magnet that holds a note on a refrigerator.

The strange invisible force that attracts pieces of iron and steel to a magnet is still not fully understood.

A magnet affects only certain materials, and then only when they are close to it. They must be within its magnetic field.

M agnetism produces electricity and electricity produces magnetism.

The easiest way to show how electricity makes magnetism is to find out how magnets are made. Suppose you wanted to make a horseshoe magnet; you would take a piece of steel and wind some fine copper wire around it, starting on one end of the horseshoe and winding the wire around the steel until it came to the other end.

Connect a battery to each end of the wire, a current of electricity will flow through the wire having such an influence on the steel, that it is converted into a magnet. When the current is shut off and the wire removed you will have a permanent magnet.

Think of magnetic materials as either hard or soft. Soft materials are used in devices where a change in the magnetic field is necessary in the operation of the device. Sometimes a very rapid change is required. These are called electromagnets. An electromagnet can be activated by a switch; the magnetic field can be turned on and off by a switching action.

Planet earth is a huge magnet and surrounding the earth is the magnetic field produced by earth's magnetism. When a conductor carries an electrical current, it sets up a magnetic field around the conductor.

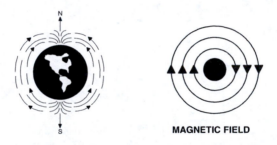

MAGNETIC FIELD

An electromagnet is a convenient way of turning electrical energy into rotary motion, using the forces of magnetic attraction and repulsion in electric motors.

If you would do the same thing with a horseshoe made of soft iron, instead of steel, it would not be a magnet after you stopped the current of electricity from going through the wires, although the piece of iron would be a strong magnet while electricity was going through the wire around it.

Magnetic induction is a very *temporary* phenomenon. When the paper clip that touches the magnet is pulled away from the magnet, it loses its power to hold up the other paper clips. It's like electricity; when the plug is pulled, the wire loses it's charge.

The steel magnet is called a permanent magnet and its ends, or "poles", are named North and South. The strongest point of the magnet is at the poles, while at the point marked +, there is hardly any magnetism.

The magnets made of iron are called electromagnets, because they exhibit magnetism only when the current of electricity is flowing around them.

If you present the North pole of a magnet to the South pole of another magnet, they will atrract and hold fast to each other, but if you present a North pole to a North pole or a South pole to a South pole, they will repel each other and there will be no attraction.

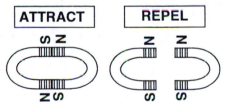

If you take a steel magnet and put it under a piece of paper and sprinkle steel shavings on the paper, you will see the steel shavings arrange themselves in a pattern. Each steel shaving becomes a magnet and would have North and South poles just as the large magnet under the paper has.

Every conductor carrying current has a magnetic field surrounding it.

When a conductor carrying current is passed through a plane, a number of iron filings will arrange themselves in circles or rings to show that the magnetic effect is circular and at right angles to the conductor. That is, if the current flows in a vertical direction, the magnetic field extends outward in a horizontal direction. If the current is reversed, the iron filings will reverse their position also.

We think of magnetism in terms of lines of force although actually, this may not be the case. Magnetism can be thought of as a fluid, but in order to think of it in terms of quantity, it's best to think of magnetism in the form of lines or *rubber bands*.

L ines of force act much like rubber bands for they try to shorten themselves as much as possible and in doing so the lines become thicker or more numerous.

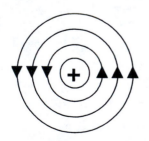

This view is looking at the end of a conductor carrying current away from us. The direction of the lines of force is counter-clockwise.

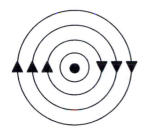

This view shows the current flowing towards us and the lines of force are reversed to a clockwise direction.

I f the lines of force were visible they would appear in circles having a common center around the conductor carrying current.

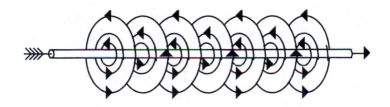

If two conductors carrying current in the same direction are placed side by side, the lines will form an oval path except near the conductors where a few lines will surround the individual conductors.

Since the lines of force are going in the same direction their magnetic fields will aid each other. The lines of force join and form loops around both wires and make a stronger magnet.

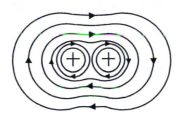

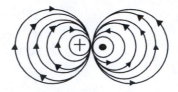

If two conductors carrying current in the opposite direction are placed side by side, their magnetic fields will oppose each other, since the lines of force are going in opposite directions. The lines of force cannot cross, and the fields move the wires apart.

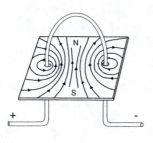

You have seen how the lines of force look around a straight conductor; now look at the loop. The effect is the same, for one side of the loop is carrying current in one direction and the other side of the loop is carrying current in the opposite direction.

The left hand rule is used to determine the direction of magnetism when the direction of current is known. This rule is based on the electron theory of current flow (from negative to positive) and is used to determine the direction of the lines of force in an electromagnetic field.

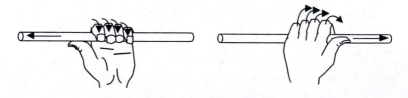

If the thumb of the left hand indicates the direction of current, the extended fingers will indicate the direction of the lines of force around the conductor.

The direction of an induced current is known as Lenz's law, which states: An induced current has such a direction that its magnetic action tends to resist the motion by which it is produced.

It is important for the electrician to understand the meaning of inductance. Every complete circuit has some inductance. Even the simple complete loop circuit or a single-turn coil has inductance. The coil having many turns has a higher inductance than a coil having fewer turns.

The picture to the left shows the relationship of induction to water.
When a garden hose is wrapped around a post several times, the coiling will *oppose the normal flow* of water to slow it down. A plumber refers to this as back pressure, like a restriction to the flow.

E lectrically it offers a resistance to the normal flow of current. We call it *inductance*.

Inductance uses electrical energy to create a magnetic field and the magnetic field restores the energy back to the line when it collapses.

For a DC circuit, inductance affects the current flow when the circuit is turned on or off. When the switch is turned on, current flows through the circuit and the lines of magnetic force expand outward around the circuit conductors and the current rises from zero to its maximum value. Whenever a current flow changes, the induced magnetic field changes and opposes the change in current whether it be an increase or decrease and inductance will slow down the rate at which the change occurs. When the switch is opened in a DC circuit, the current will drop very rapidly towards zero causing the magnetic field to collapse and generate a very high voltage, which not only opposes the change in current but can also cause an arc across the switch.

The big difference is an AC circuit is constantly switching on and off, reversing direction 60 times a second. So circuit *inductance affects AC circuits all the time.*

In electrical formulas, the letter "L" is the symbol letter used to designate inductance. The unit of measure for inductance is the *henry*. Because a coil of wire has more inductance than a straight wire it is called an inductor.

Remember, in an inductive circuit when current increases, the circuit stores energy in the magnetic field. When current decreases, the circuit gives up energy from the magnetic field. In an AC circuit, the magnetic field is always changing. Every circuit has some inductance, although it may be so small that its effect is negligible.

If wire can be bunched together in the form of a coil with an iron core we have what is known as an electromagnet.

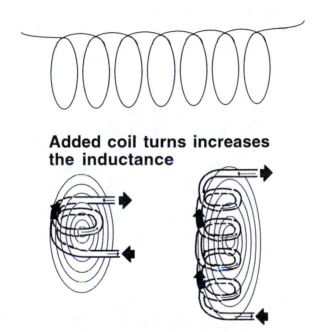

Added coil turns increases the inductance

The strength of the magnetic field depends on the amount of current flow. If the conductor is straight there will be little self-induction, but if coiled, the magnetic effect is greater.

The strength of a magnet also depends on the resistance of the magnetic path. This resistance we call *reluctance* instead, to distinguish it from resistance in an electrical circuit. The greater the reluctance (resistance), the less strength of the magnet.

In an electromagnet circuit there is a pressure which maintains the magnetic lines in the circuit causing them to overcome the reluctance. The unit for magnetic pressure is the ampere-turn. The number of ampere-turns always being equal to the number of amperes flowing through the coil multiplied by the number of turns.

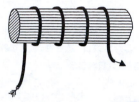

This coil has four turns of wire carrying 2 amperes. The number of ampere turns is equal to 4 x 2 = 8 ampere-turns.

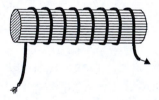

This coil has eight turns of wire carrying 1 ampere. The number of ampere turns is equal to 8 x 1 = 8 ampere-turns.

A magnetic circuit behaves very much the same as an electrical circuit. Ampere turns can be compared with volts, lines of force with current, and reluctance with resistance.

In today's electrical world you will find many applications of electromagetism. To name a few would include, coils to relays, motor starters, power contactors, solenoids, elevator braking, lifting magnets, magnetic clutches, etc.

Relays, solenoids, etc. contain a magnetic coil when energized by an electrical current passing through it causes the iron armature to move in the frame.

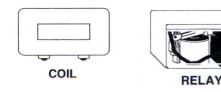

COIL   **RELAY**

SOLENOID

It would *not* be an exaggeration to say that all of the electricity used in the industry is based on the generation and utilization of *magnetic fields*.

Around 1820 Joseph Henry discovered that when a wire is charged with electricity it reacted within itself as the connection is broken. This is called self-induction. In respect for his scientific achievements, a unit of measurement of mutual induction is called the *henry*.

The father of electricity is Michael Faraday, who in 1831 demonstrated electromagnetic induction manipulating magnetism to make electricity.

Now let's examine how magnetism makes electricity by electromagnetic induction.

The definition of induction is "the process by which an electrical conductor becomes electrified when near a charged body, by which a magetizable body becomes magnetized when in a magnetic field or in the magnetic flux set up by a magnetomotive force, or by which an electromotive force is produced in a circuit by varying the magnetic field linked with the circuit".

It has been found that the influence of a magnet is very strong at its poles, and that this influence is always in the same lines. This influence has been described as "lines of force". Of course the lines are *invisible*, but they are always present. The meaning of this term "lines of force" is used to designate the strength of the magnet.

Many years ago, the scientist Faraday discovered that by passing a closed loop of wire through the magnetic lines of force existing between the poles of a magnet, the magnetism produced the peculiar effect of creating a current of electricity in the wire.

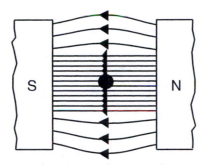

The conductor is shown by the large black dot, which is the end view of the conductor. This conductor is lying in the magnetic field, but this conductor is *not* carrying any current. Note; at this point the magnetic field is not distorted.

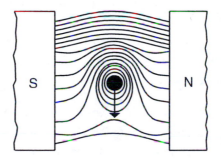

Now the same conductor is placed into motion, and as you can see the magnetic field is distorted. Now the magnetic lines or lines of force are in their mechanical action, something like rubber bands. An electrical pressure (magnetoelectricity) is generated in the conductor and this pressure is capable of moving a current in the conductor if its circuit is complete. The electric current produced by moving a conductor in a magnetic field is called an induced current.

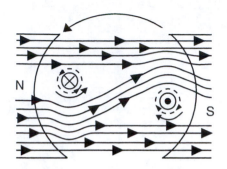

T he current is only induced when the conductor is moving, and that the direction of the current is reversed when the motion changes direction. If a sensitive voltmeter is connected to the conductor, the voltmeter would record in one direction when the conductor is moved down and in the opposite direction when the conductor is moved up through the field. If a conductor is moved up and down through a magnetic field a voltage will be induced in it.

There are three fundamental conditions which must exist before a voltage can be produced by magnetism.

I. There must be a *conductor* in which the voltage will be produced.

II. There must be a *magnetic field* in the conductor's vicinity.

III. There must be a relative *motion* between the field and the conductor. The conductor must be moved so as to cut across the magnetic lines of force, or the field must be moved so that the lines of force are cut by the conductor.

When a conductor or conductors *move across* a magnetic field so as to cut the lines of force, electrons *within the conductor* are impelled in one direction or another. Thus, a voltage (force) is created. This is called a *generator.*

A generator converts *mechanical* energy into *electrical* energy. Whereas an electric motor is just the opposite, it uses electrical energy to perform a mechanical function. An example would be a fan.

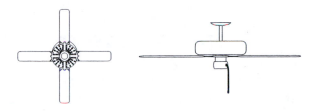

T he water falling from a dam is the mechanical energy used to drive a generator to produce electrical energy.

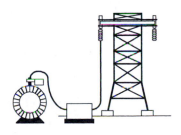

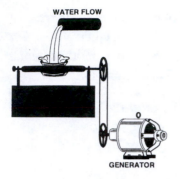

WATER FLOW

GENERATOR

Whatever the original source of energy - water, coal, oil, gas, steam, the sun, the wind - the final step is always the conversion of *mechanical energy* from rotation of a generator to produce *electrical energy*.

DC GENERATOR

To provide a constant motion, the conductor would need to move up and down continuously. The practical way of motion is to have the conductor travel in a circular motion. This is the principle of the *generator* (dynamo); the spinning of the magnets.

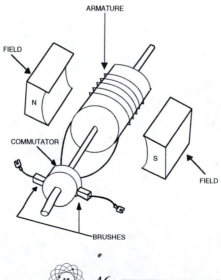

ARMATURE

FIELD

N

COMMUTATOR

S

FIELD

BRUSHES

When a conductor is *moved* through a magnetic field in such a way that it cuts the lines of magnetic flux, a force is applied to make *electrons move*. This is the basic principle of how a generator works.

S hown below is the simplest form of a generator. It consists of a single loop of wire, which is placed between the poles of a permanent magnet and made to *rotate*. As the loop of wire rotates, it cuts through the magnetic lines of force and a *voltage is developed*.

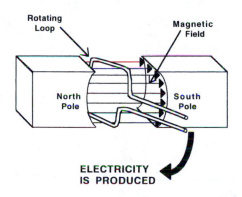

All generators, whether AC or DC consist of a rotating part and a stationary part. The rotating part of a DC generator is referred to as the *armature*. The coils that generate the magnetic field are mounted on the stationary part which is referred to as the *field*. In most AC generators the opposite is true, the field is mounted on the rotating part referred to as the *rotor* and the armature is wound on the stationary part referred to as the *stator*.

In carrying this discovery into practice in the making of generators we use copper wire instead of iron. Copper has less resistance than iron to the flow of current.

The complete armature, the iron core, winding, commutator and shaft, is positioned inside an iron frame or housing. The field poles are made of iron, either solid or laminations and support coils of wire called field windings. The field winding is an electromagnet.

ARMATURE

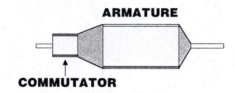

COMMUTATOR

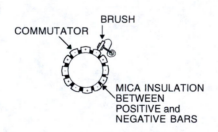

COMMUTATOR

BRUSH

MICA INSULATION
BETWEEN
POSITIVE and
NEGATIVE BARS

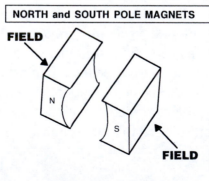

NORTH and SOUTH POLE MAGNETS

FIELD

N

S

FIELD

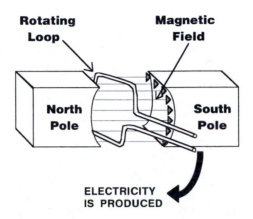

When the armature revolves through the lines of flux, the magnetic energy forces current to flow in the wire. When the wire in the armature goes *down* the field current flows in *one direction*; but when the wire goes *up* the field, the current flows in the *other* direction.

When a closed loop is passed up and down between the poles of a strong magnet there is an opposition felt to the passage of the wire back and forth.

This is due to the influence of magnetism upon the current produced in the wire as it cuts through the lines of force, and inasmuch as these lines of force are always present at the poles of a magnet, you will see that no matter how many times you pass the loop of wire up and down, there will be created in it a current of electricity by its passage through the lines of force.

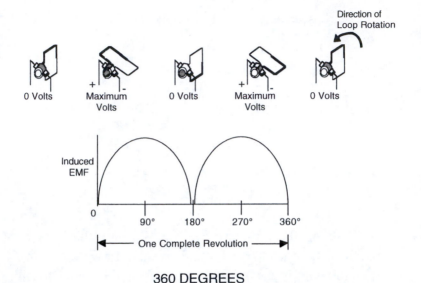

Direction of
Loop Rotation

0 Volts   Maximum Volts   0 Volts   Maximum Volts   0 Volts

Induced EMF

0   90°   180°   270°   360°

One Complete Revolution

360 DEGREES

The brushes are positioned on *opposite* sides of the commutator so that they pass from one commutation half to the other at the instant the loop reaches the point in its rotation where the induced voltage reverses polarity. The brushes are effectively shorting the ends of the loop directly together. So instead of the output voltage reversing polarity after one-half revolution, the voltage output for the second half revolution is identical to that of the first half.

In a practical generator, the revolving loop of wire in the armature will contain several loops of wire cutting the magnetic flux of the fields.

When a coil or armature makes one complete revolution, it passes through 360 *mechanical degrees*, when an emf current passes through one cycle, it passes through 360 *electrical time degrees. (360°)*

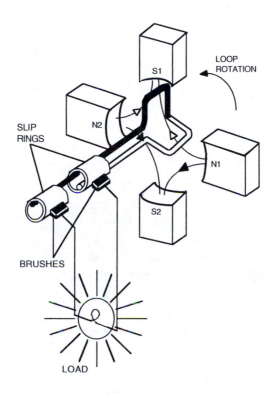

If the generator makes two complete revolutions per second, the output frequency will be *two Hz* (cycles). In other words, the frequency of a two-pole generator happens to be the same as the number of revolutions (cycles) per second. As the speed is increased, the frequency is increased.

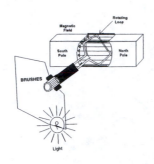

All we have to do then, is to collect this electricity from the two ends of the wire and use it. If we should attach two wires to the two ends of this wire on the spool, they would be broken off when it turned around, so we must use another method. The wires are placed on the end of the spool (armature) so they will not touch each other, and fasten these ends to a *"commutator"*.

Brushes attach to the commutator to collect the electricity being produced and send it to the circuit.

The constant revolving of the armature creates a continuous electricity that is driven from the wire through the commutator to the brushes to the light.

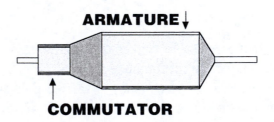

ARMATURE ↓

COMMUTATOR

The commutator switches the wires outside the generator while the armature turns to keep the current flow in the same direction. This is called direct current (DC). If a commutator is not used, the current coming out of the generator will change directions as the loops revolve. This is called alternating current (AC).

You can change the amount of electricity that is produced by changing the strength of the magnetic field or by cutting more lines of force with the conductor in a shorter length of time.

To increase the amount of electricity produced by a moving conductor you could increase the length of the wire that passes through the magnetic field, use a stronger magnet, or move the conductor faster through the field. The length of the conductor can be increased by winding it in several turns to form a coil.

S uppose, that instead of using one single loop of copper wire, you wound upon a spool a long piece of wire and turned the spool rapidly between the poles of the magnet. You would be cutting the lines of force by the same wire many times, and every time one length of wire cut through the lines of force some electricity would be generated in it, and this would continue as long as the spool was revolved. But as each length would only be part of the one piece of wire, you will easily see that there would be a great deal of electricity generated in the whole piece of wire.

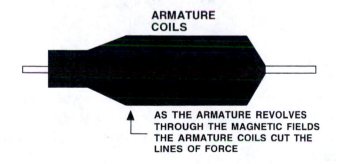

ARMATURE COILS

AS THE ARMATURE REVOLVES THROUGH THE MAGNETIC FIELDS THE ARMATURE COILS CUT THE LINES OF FORCE

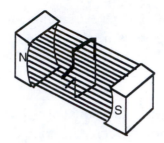

The value of an induced voltage in a conductor will depend on the rate at which the lines of force are cut. If a conductor crosses a field of 100 million lines of force in one second of time, one volt will be induced in the conductor. If the conductor crosses a field of 50 million lines of force, and at the same speed, 1/2 volt will be induced in the conductor. But if the conductor crosses the field in 1/2 second, one volt will be induced. The voltage induced in a conductor depends on the speed of the moving conductor and the strength of the magnetic field (greater strength, more lines of force).

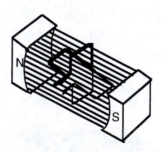

If two conductors are cutting across the same lines of force the voltage would be twice as great as when only one conductor is used. This is the reason an armature is built with several coils of wire, to cut the magnetic lines of force and increase the voltage.

When a conductor is forced straight through a magnetic field or at right angles the voltage induced will be greater than if the conductor moves through the field diagonally at the same speed.

If the magnetic field is 1" thick the conductor shown in the top view cuts all the lines. If the conductor moves diagonally instead of straight through as shown in the lower view it will only cut about 70% of the lines.

Example: If cutting the lines straight induced 100 volts in the conductor, cutting the lines diagonally would result in only 70 volts.

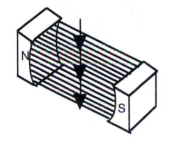
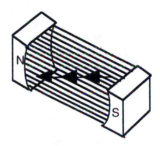

A *dynamo* is an electric machine used to convert mechanical energy into electrical energy, and is called a generator or alternator.

The dynamo must be turned mechanically to produce the electricity. The earliest forms of mechanical energy used falling water. The water turned a turbine which rotated the armature through the fields.

AC ALTERNATOR

For a single loop rotating in a two-pole field (one north pole and one south pole), you can see that each time the loop makes one complete revolution the current *reverses direction twice*. A single hertz (cycle) will result if the loop makes one revolution each *second*. A complete cycle is 360°. There are *two alternations* in one complete cycle. One positive alternation and one negative alternation. This is called a *sine wave*. By reversing the direction twice in one cycle this is called alternating current (AC).

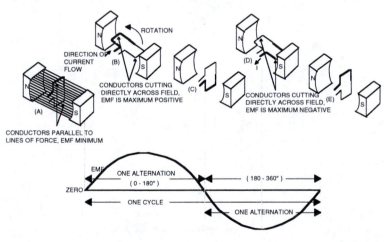

THE AC SINE WAVE

To convert AC to DC a switch must be operated twice for every cycle. If the generator output is alternating at 60 Hz (cycles), the switch must be operated 120 times per second to convert AC to DC. Obviously, it would be impossible to operate a switch manually at this high rate of speed.

A direct current generator (DC) uses a *commutator* as a switch to change the alternating current to direct current.

The carbon brush, as it slides on the revolving commutator, reverses the connections of the connection in the armature to the external circuit at the instant when the voltage of the conductors is zero and changing in direction. The commutator switches the wires outside the generator while the armature turns, thus keeping the current flow in the *same direction* at all times. If a commutator is not used, the current coming out of the generator will *change direction* as the armature turns.

In building construction we use alternating current (AC) rather than DC. Alternating current changes direction at regular intervals of time. A 60 cycle alternating current changes direction 120 times every second; 60 positive and 60 negative alternations every second.

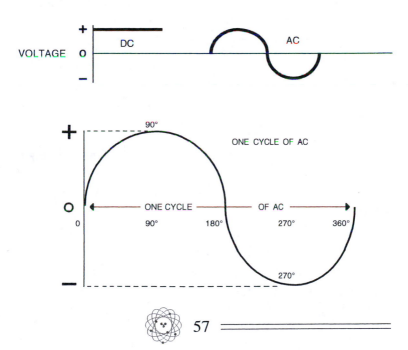

The comparison of DC and AC is shown below.

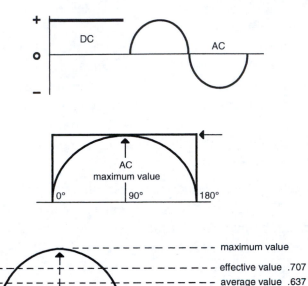

T here are two maximum or peak values for each complete cycle of AC. Usually AC voltages and currents are expressed in root mean square (RMS) or effective values.

When DC or AC flows through a resistance, electric energy is converted into heat. AC varies continuously between maximum values and zero and is lower than DC.

The AC circuit will have to be increased to 1.414 amperes before it will produce the same heating effect as will one ampere of DC current. Similarly, the peak voltage is 1.414 times the RMS voltage.

RMS VALUE = MAXIMUM x .707
MAXIMUM VALUE = RMS/.707

Peak Maximum

RMS Effective Value .707

METER READS
RMS EFFECTIVE VALUE

The RMS effective value is the same as the DC maximum value. RMS is .707 (or approximately 70%) of the AC maximum peak. AC maximum peak is an instantaneous value changing 60 times a second. The AC alternator is producing a sinusoidal (sine curve) waveform.

Alternating voltage or current changes continuously with time. It rises from zero to a maximum value in one direction and decreases back to zero. It then rises to the same maximum value in the opposite direction and again decreases to zero. These values are repeated again and again at equal intervals of time.

Root-mean-square current is the abbreviated form of "the square root of the mean of the square of the instantaneous currents".

B y the AC reversing 60 times a second, the volt or ammeter is *not* reading the peak to peak (maximum), the meter is reading the RMS (effective value) which is .707 or 70% of maximum.

METER READS
RMS EFFECTIVE VALUE

The RMS value is used to compare the AC heating effects to the DC heating effects.

The effective value (RMS) of an AC voltage or current is the value that will cause the same amount of heat to be produced in a circuit containing only resistance that would be caused by a DC voltage or current of the same value.

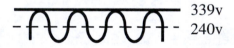

339v
240v

If the RMS (effective value) is 240 volts, the maximum (peak) voltage of an AC system is 339v.

240v/.707 = *339 volts*

240 volts of DC will produce the same effect as 339 volts AC.

METER READS

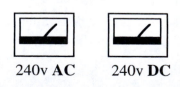

240v **AC**   240v **DC**

Example: If the maximum value of an AC current is 50 amps, the RMS value would be approximately _____ amps.

(a) 25   (b) 30   (c) 35   (d) 50

Solution: 50 amps x .707 = 35.35 or *35 amps.* The ammeter would read 35amps.

Two lamps, one is 6 volts (RMS) supplied from an AC transformer, the other lamp is a 6 volt DC supplied from a battery. The identical lamps will have the same brightness showing the effect of the two voltages are the same and the power (watts) in both circuits are the same. •Remember, the AC is 6 volts (RMS), the maximum (peak) voltage would be 6v/.707 = 8.48 volts to give this same effect.

There are only *two types* of generators, AC and DC. Over 90% of all electric power is AC.

AC generators do not have commutators and this makes them far superior to a DC generator. AC generators are also called *alternators*, since they produce an *alternating current.*

AC generators can be built with much larger power and voltage ratings than DC generators. The reason is that the AC generator output connections are bolted directly to the stationary windings.

All generators operate on the same basic principle; a magnetic field cutting through conductors or conductors passing through a magnetic field.

There are two groups of conductors:

I. A group of conductors in which the output voltage is generated.

II. A group of conductors through which direct current (DC) is passed to obtain an electromagnetic field of fixed polarity (excitation).

The conductors in which output voltage is generated are referred to as *armature windings*. The conductors in which the electromagnetic field originates are referred to as the *field windings*.

There must be a *motion* between the armature windings and the field windings. AC generators are built in two major assemblies, the *stator* and the *rotor*.

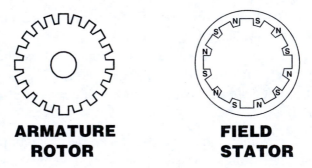

ARMATURE
ROTOR

FIELD
STATOR

There are two types of motion, either the revolving armature (rotor) or the revolving field (stator).

In the *revolving armature* AC generator, the stator provides a stationary electromagnetic field. The rotor acting as the armature, revolves in the field, cutting the lines of force, producing the desired voltage. In this generator, the armature output is taken through slip rings and thus retains its AC characteristic.

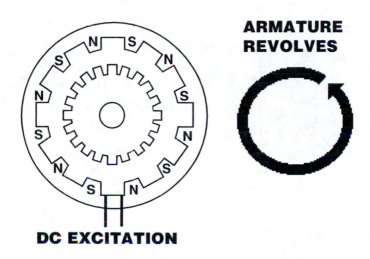

ARMATURE REVOLVES

DC EXCITATION

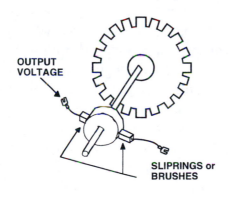

OUTPUT VOLTAGE

SLIPRINGS or BRUSHES

The revolving armature AC generator is *seldom* used. Its primary limitation is the fact that its *output* power is conducted through sliding contacts, sliprings, and brushes. The sliprings and brushes *limit the amount of voltage* that can be carried through them due to arcing and flashovers. Consequently, revolving armature AC generators are limited to low-power, low-voltage applications. An example would be an automobile alternator.

Remember you can just as easily rotate the *magnet assembly* (the fields). This is what the Yugoslavian Nikola Tesla developed and received patents on in 1888.

The *revolving field* AC generator is by far the most widely used today. In this type of generator, direct current from a separate source (excitation) is passed through windings on the rotor by means of sliprings and brushes. This maintains a rotating electromagnetic field of fixed polarity. The rotating magnetic field cuts through the armature windings imbedded in the surrounding stator. As the rotor turns, AC voltages are induced in the windings since magnetic fields of first one polarity and then another cut through them. Now here is the important part; since the output power is taken from *stationary windings,* the output may be connected through *fixed* terminals and not revolving sliprings or brushes that would limit high voltages. Sliprings and brushes are adequate for the DC field (excitation) supply because the power level in the field is much smaller than in the armature circuit.

**FIELD
ROTOR**

**ARMATURE
STATOR**

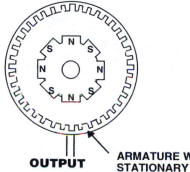

FIELD REVOLVES

OUTPUT
VOLTAGE

ARMATURE WINDINGS
STATIONARY

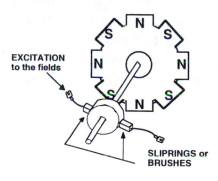

EXCITATION
to the fields

SLIPRINGS or
BRUSHES

Excitation through the rotating fields is provided at lower voltages.

AC - DC GENERATOR SUMMARY

DC voltage changes are obtained by using series resistors which causes *low efficiency* due to heat loss.

DC generator ratings are limited to relatively low voltage and power values as compared to AC generators.

AC armature stator voltages of 13,800 are common compared to 750 volts for a large DC generator.

The initial cost and the maintenance and repair costs for AC is considerably less than the costs for DC.

Although there are a number of applications where *DC* either must be used or will do the job better than AC, such as:

• Charging of storage batteries
• Electronics
• Electroplating process
• Excitation of the field windings of generators
• Variable speed motors

There are special jobs that require *heavy starting torque* and *high rate of acceleration* such as locomotives and monorail trains which are driven by traction motors which require DC. Using DC motors in these applications eliminates the need for clutches, gear shifting transmissions, differential gearing, drive shafts, and universal joints.

AC is changed to DC by *rectifiers* or *motor-generator* sets. Thus, the costly conversion to DC is needed only for certain applications.

THE MOTOR

A machine that converts mechanical energy into electrical energy is called a *generator*.

A machine that converts electrical energy into mechanical energy is called a *motor*.

Otto Von Guericke, in 1660, built the first electric generator. The machine consisted of a glass ball containing sulfur mounted on a shaft with a hand crank. When the glass ball was cranked to a high speed, a cloth was applied, and sparks would jump between a spark gap connected to brushes that touched the rotating glass ball.

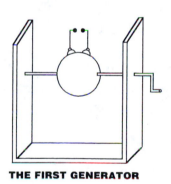

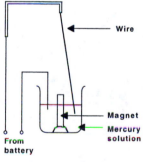

THE FIRST GENERATOR   **THE FIRST MOTOR**

Michael Faraday, in 1821, came up with an apparatus to prove that electrical energy could be converted to mechanical energy through *magnetism*. It occurred to Faraday that if electricity could produce magnetism, couldn't magnetism produce electricity? Faraday invented the electromagnetic motor.

The early hand cranked generators operated by moving a coil back and forth in front of a permanent magnet or by moving a magnet inside of a coil in a similar manner. Today we use a rotary motion between the coil and the magnet instead of the back and forth motion.

In 1832 Hippolyte Pixii built the first machine that had fixed coils wound on a U-shaped iron rod. Above the rod a horseshoe magnet was attached to a shaft turned by a hand crank. When the hand crank was turned, the rotating magnet produced an *alternating current* in the coils. However, at that time there was *no use for an alternating current.*

Andre Ampere worked with Pixii to improve the machine by installing a cam operated switch designed to reverse the alternations and produce a more usable current close to that of a battery.

This cam operated switch is a mechanical switch that maintains current in one direction; *direct current*. In today's generators this *commutator* is the action of the brushes riding on the bars of the armature.

A DC generator may be operated as a DC motor, or a DC motor may be operated as a generator. The two machines are structurally identical. The generator, like the motor, consists of an electromagnet, an armature, and a commutator with brushes.

A motor must develop a continuous rotary motion. This basic twisting force is called torque.

The turning force on the armature depends on several factors, including field strength, armature current strength and physical construction of the armature. The larger the armature the greater the *torque*.

The torque determines the energy available for doing the work. If a motor does not have enough torque for the load, it will stall.

The lines of force coming from the North pole become compacted beneath the left side of the conductor while they become thinned out as they continue on the right side of the conductor. Since there are fewer lines above the left side and more lines concentrated above the right side. The left side is both pushed and attracted up at the same time the right side is forced down. This combination of forces create torque that turns the armature of a motor.

As the conductor cuts through the magnetic lines producing the torque to rotate the motor it acts like the force that the wing of an airplane feels as it lifts. The top of the wing feels a lower pressure than the bottom. This lower pressure is what causes the wing to create lift, enabling the plane to become airborne.

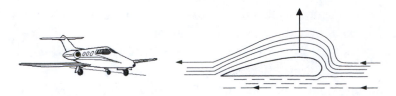

Motors operate from the principle that electrical energy can be converted to mechanical energy. For generators, the principle is simply reversed.

The dynamo (generator) can be changed to a motor by applying the electricity to the brushes which goes through the commutator into the armature and round the magnet, and so create the lines of force at the poles and magnetize the iron of the armature. Each segment of the armature becomes a North and South pole. The effect it has is the North pole of the field magnet is repelling (forcing away) the North pole of the armature and at the same time drawing towards itself the South pole of the armature. In the mean time, the South pole of the field magnet is repelling (forcing away) the South pole of the armature and at the same time drawing towards itself the North pole of the armature.

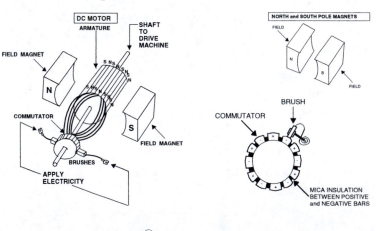

T his makes the armature turn around and the same poles are again presented to the field magnet, when they are acted upon in the same manner, which makes the armature revolve again, and this action continues as long as electricity is brought through the wires to the brushes. Thus, the armature turns around with great speed and strength, and will drive a machine to which it is attached.

The speed and strength of the motor are regulated by the amount of iron and wire upon it, and by the voltage supplied to the brushes. Motors are made in many different forms, but the principle is practically the same.

The functions of a DC generator and a DC motor are interchangeable in that a generator may be used operated as a motor, and a DC motor can be used as a generator.

TRANSFORMER

The word transform means to "change in form".

In 1831 Michael Faraday constructed what was basically a transformer. The action of one coil *inducing* (induction) an effect on another coil.

A transformer is a basic and very useful device, so widely used you would rarely walk more than 200 feet to find one.

A transformer *does not* generate electrical power. It *transfers* electrical power. A transformer is a voltage changer.

A device, usually consisting of two insulated windings on a common iron core, in which alternating current is supplied to one winding and by electromagnetic induction which induces alternating EMF'S in the other winding.

One of the windings is designated as the *primary* and the other winding as the *secondary*. The primary winding *receives* the energy and is called the *input*. The secondary winding *discharges* the energy and is called the *output*.

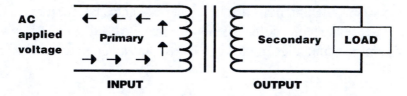

The transformer works on the principle that energy can be efficiently transferred by magnetic induction from one winding to another winding by a varying magnetic flux produced by *alternating current.*

An electrical voltage can only be *induced* while there is a relative *motion* between a wire or a circuit, and a magnetic field. *Alternating current* provides the motion required by *changing direction 60 times a second.*

Direct current (DC) is not transformed, as DC does not vary in its amount from one second to the next. This is the reason we wire buildings with AC, we can't transform DC.

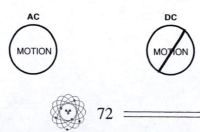

The two windings, primary and secondary, are *linked* together with a magnetic circuit which must be common to both windings. The *link* connecting the two windings, in a magnetic circuit, is the *iron core* of which both windings are wound. Iron is an extremely good conductor for *magnetic* fields. The core is not a solid bar of steel, but is constructed of many layers of thin steel called *laminations*.

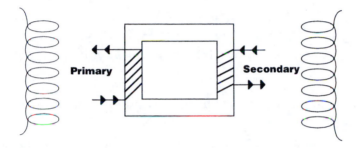

S ince the primary and secondary are wound on the same iron core, when the primary winding is energized by an AC source, an alternating magnetic field called *"flux"* is established in the transformer core. The flux surrounds both the primary and secondary windings and *induces* a voltage in them.

Since the same flux cuts both the primary and secondary windings, the same voltage is induced in *each turn* of each winding.

The voltage in the primary will be the same as the supply voltage, but the secondary voltage will depend on the number of *turns* in the secondary winding in proportion to the number of *turns* in the primary winding.

If the secondary has the same number of turns as the primary, the voltage will be the same in both windings.

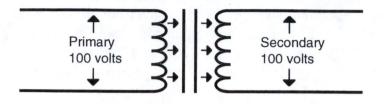

1 to 1 ratio
Primary = 100 turns = 100 volts
Secondary = 100 turns = 100 volts

By changing the *ratio* of turns in windings you change the voltage.

If the secondary has *half* the number of turns as the primary, the secondary voltage will be *half* as high as the primary voltage.

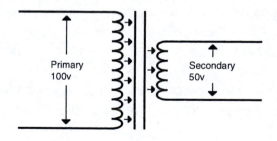

Voltage is stepped down
at a **2 to 1 ratio**

Primary = 100 turns = 100 volts
Secondary = 50 turns = 50 volts

If the secondary has *twice* the number of turns as the primary, the secondary voltage will be *twice* as high as the primary voltage.

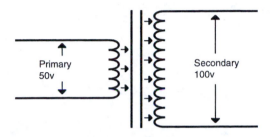

Voltage is stepped up
at a **1 to 2 ratio**

Primary = 50 turns = 50 volts
Secondary = 100 turns = 100 volts

The voltage is *stepped down* when the primary has more turns than the secondary. The voltage is *stepped up* when the secondary has more turns than the primary.

When the primary winding and the secondary winding have the *same* number of turns, there is no change in voltage, the ratio is 1/1 or *unity*.

The ratio between the voltage and the number of turns on the primary and secondary windings is called the *turns ratio*.

It is customary to specify the ratio of transformation by writing the primary (input) number first.

Example: 30 to 1 is a step-down transformer, whereas a 1 to 30 would be a step-up transformer.

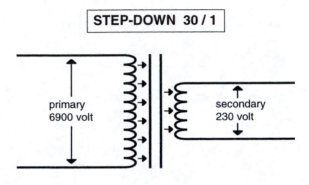

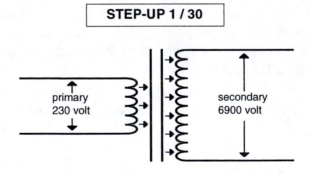

I like to think of the secondary winding as a baseball catcher's mitt. As the alternating current reverses direction through the primary winding, an electron is induced into the invisible magnetic field and the catcher's mitt (secondary winding) catches these electrons.

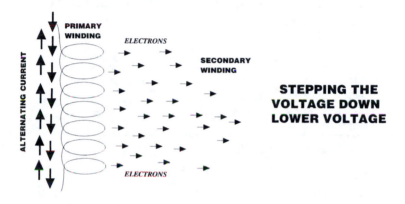

The size of the catcher's mitt (secondary winding) will determine the voltage output of the secondary. With a smaller mitt less electrons will be caught, with a larger mitt more electrons will be caught.

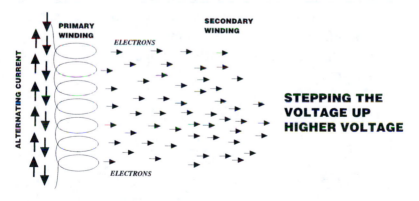

The following will show why we use AC alternating current rather than DC direct current.

The sketch below shows a DC power plant serving the customer located 1000 feet from the source with a 240 volt DC 200 amp load. The wire from the power plant to the customer (1000' in distance) would be sized to carry 200 amps.

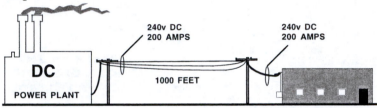

The sketch below shows an AC power plant serving the same customer with a 200 amp load.

The difference is very clear. The power plant now produces 2400 volts instead of 240 volts. With a 10/1 ratio *transformer* located at the site of the customer the 2400 volts is reduced to 240 volts.

The important advantage is the amperes are reduced from 200 to 20 on the transmission lines of 1000 feet. This is a huge savings in costs as we size wire to amperes. Now instead of a wire sized for 200 amps, the wire is sized for 20 amps. The weight factor on the supporting poles is also reduced.

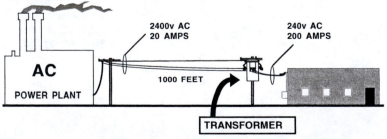

If DC current is required by the customer, AC would be installed to the customers building and then a *rectifier* would be used to change the AC to DC.

By using AC there will be less voltage drop and power loss.

The only respect in which alternating currents (AC) differ from direct currents (DC) is the manner in which they flow through a circuit. A direct current continually flows in the same direction and is practically constant in value. But an alternating current, as the name suggests, alternates in both its direction and its value. In other words, it is constantly and regularly reversing its direction of flow, and between the times of reversal its value increases from zero to a maximum value, and then decreases to zero again. When it reaches zero, it reverses and again increases to the same maximum value but in the opposite direction. An alternating current can thus be defined as one that reverses its direction of flow at regular short intervals. A complete set of values through which an alternating current passes is known as a *cycle*, and the number of these cycles through which the current passes in *one second* is known as the *frequency*.

The same laws apply to AC as to DC and they are measured in the same units (volts, amps, ohms). However, due to their periodic changes, certain effects are present in AC circuits that are not found in DC circuits, and consequently their study is somewhat more complicated although not much harder.

We use AC in the wiring of buildings rather than DC. AC gives a more flexible service and can be transmitted with less line loss over smaller wires because of the ease with which the voltage can be raised or lowered by means of a *transformer*.

You have already learned that whenever a wire crosses magnetic lines of force a current of electricity will flow in the wire if the wire forms a closed circuit. Also you have learned that wherever there is a current of electricity flowing there is a magnetic field of some form present. If you see these two points clearly, you should have no difficulty in understanding *self-induction*. When a wire moves across a magnetic field there is always an electromotive force induced no matter what the source of the magnetic field. The field may be produced by an electromagnet or a number of horseshoe magnets and the wire may move across the north and south poles alternately. This method is the alternating current generator.

But an electromotive force may be induced in a coil of wire by its own magnetic field provided the magnetic field changes. This is called *self-induction*. Self-induction is purely a matter of *magnetic action*, and that is essentially the same thing as ordinary induction.

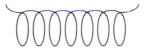

To understand an induction coil, you must first understand a *capacitor*, which is a necessary part of such a coil. A capacitor is made up of two sets of conducting plates, insulated from each other by a substance called a *dielectric* (insulator).

Examples of dielectrics are mica, glass, air, wax, etc. A capacitor has the property of absorbing an electric charge on its dielectric and discharging it again when the two sets of conducting plates are connected together so as to complete the circuit between the conducting plates.

CAPACITOR

The sketch shows the construction of a capacitor.

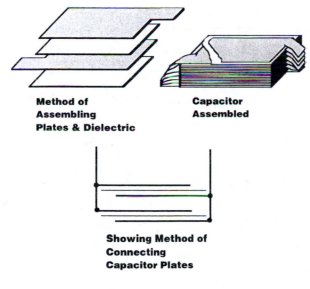

**Method of
Assembling
Plates & Dielectric**

**Capacitor
Assembled**

**Showing Method of
Connecting
Capacitor Plates**

The plates are alternately assembled between the sheets of dielectric; the plates themselves usually being made of copper or lead foil. First one plate is laid flat and covered with a sheet of dielectric, usually waxed paper or mica, then another plate is laid over the dielectric, but with its connecting tab on the opposite side. The first, third, fifth plates, and so on are connected together at their tabs by a rivet and the second, fourth, sixth plates, and so on, are connected together so that two connecting terminals result. Remember, no two plates touch each other electrically except where they are connected together at the tabs, all being separated by the sheets of dielectric. Generally commercial capacitors use waxed paper or mica as dielectrics.

If the capacitor is connected to a DC voltage source, the plates connected to the + wire will receive a + charge and the plates connected to the - side of the line will receive a - charge. These charges tend to equalize each other by sending a current around from the + plates to the - plates.

The sketch shows a capacitor as compared to water.

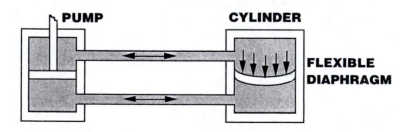

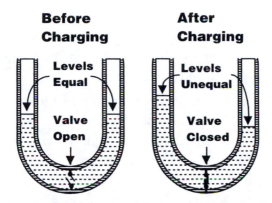

Before Charging — Levels Equal — Valve Open

After Charging — Levels Unequal — Valve Closed

W hen the plates are normal, the charge on both sets of plates is equal, but when one group of plates becomes charged more than the other group, electricity is crowded into one set of plates and removed from the other. This leaves a condition the same as the U-shaped pipes shown in the sketch where water is taken from one side and forced into the other. When the two sets of plates are left disconnected, the result is the same as having the valve closed. When the two sets of plates are connected, the charges will equalize as the water levels will equalize in the tube when the valve is opened.

The amount of electricity which can be stored in a capacitor does not depend on the thickness of the conducting plates but partly on the thickness of the dielectric and also the substance used as the dielectric. If the waxed paper or mica can be made in very thin sheets, the capacitor will store a greater charge. The area of the plates making contact with the dielectric will also effect the capacity of the capacitor. The greater the area, the greater will be the capacity to store a charge. Instead of using several large plates to get a high capacity, a number of plates are assembled in parallel.

A capacitor exists whenever an *insulating material* separates *two conductors* that have a difference of potential (voltage) between them.

Capacitors are items that are manufactured for deliberately adding capacitance to a circuit.
The electric field between the plates of a capacitor can be considered as *stored energy*, and it shows up as a *voltage* across the plates.

A capacitor blocks DC, it passes AC.

Capacitors used to be called *condensers*, so you may hear an older electrician refer to them as condensers. The meaning is the same.

An induction coil, by the action of a make and break, and a capacitor, are capable of producing very high voltages. Some coils have produced 15,000 volts when the energy was supplied by a 6 volt battery.

This increase in voltage does not represent an increase in power. In your studies of electricity always remember, it's not possible to get something for nothing. The current in the primary winding will be several amperes; whereas, the current in the secondary will be but a small fraction of an ampere. Power is lost in both of the windings as heat and power are also lost in the iron core due to the *eddy currents* and friction between the moving molecules which make up the iron.

The core of an induction coil should always be laminated or made up of pieces of soft iron wire assembled together instead of using a solid iron core. By using a core made up of a bundle of soft iron wire, each wire being insulated from the other by means of a varnish, eddy currents are reduced.

Remember that the magnetic field produced in the core of an induction coil must begin around the primary conductor. In the sketch you can see the lines of force building up around the primary conductor. They must expand into the core and also out across the secondary circuit. Since the iron core itself is an electrical conductor, current induced into it will flow in the direction of the arrows. Now eddy currents are like any other kind of currents but we give them their name because they flow in a circular path. They are, of course, undesirable. When a solid iron core is used, a complete path is provided for them and they go around through the core, *causing heat to be given off.* This is a waste of energy (heat loss).

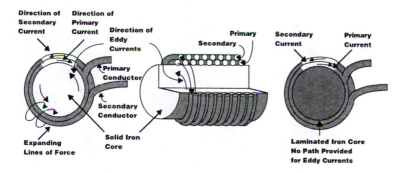

In these sketches you see that eddy currents and current induced in the secondary, flow in the opposite direction to the primary current. An induced current always flows in the opposite direction to the current in the primary conductor.

When the core is made up of soft iron bars, no path is provided for the eddy currents. Eddy currents will build up in each one of the small wires, but can only circulate in the wire in which they are induced. It is only when eddy currents are allowed to build up, as in the case of a solid iron core, that they become highly undesirable.

The difference between eddy currents and volt-amps is that eddy currents consume power as they are stored in the iron core during the cycle representing a power loss or an increase in resistance in the circuit. The volt-amp uses no power as it is stored in the magnetic field of the coil.

The dictionary defines eddy as "*a current of water running contrary to the main current*".

INDUCTANCE

C ircuit inductance is not apparent until the current in a circuit changes. In a *DC* circuit, the inductance will only effect the circuit when the switch is closed or opened. If a circuit contained only resistance the current would rise to maximum instantaneously. Inductance *opposes* the change in current. When the switch is turned on, inductance will slow down the time until it reaches maximum current. When the current reaches maximum or a steady DC value, the effect of inductance will disappear and the circuit is limited only by resistance. When the switch is opened the source voltage drops immediately to zero. The current is opposed by the inductance and takes time to drop to zero.

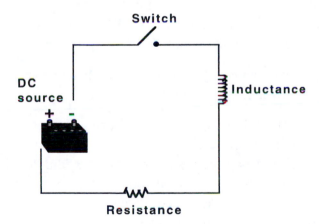

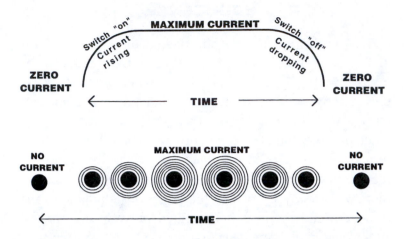

When the switch is turned "on" you can see the magnetic field slowly expanding around the conductor as current flows through the conductor. When the switch is turned "off" the magnetic lines of flux surrounding the conductor slowly collapse.

Inductance effects a DC circuit when it is turned "on" and when the circuit is turned "off".

The following sketches will show a magnetic field expanding and collapsing.

The first sketch shows two wires laid beside each other; one of which is connected to a battery and a switch, and the other to a meter. The switch is open so no current is flowing. Conductor A is the primary conductor and conductor B is the secondary conductor.

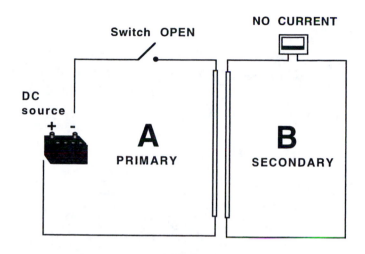

The sketch below shows the switch closed and the lines of force have expanded rapidly in primary conductor A. The lines of force grew and expanded around conductor A and cut the secondary conductor B, inducing a voltage into it.

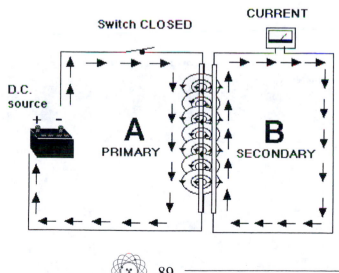

W hen the switch is opened, there will be no current
to keep the lines of force out into space and as the
field must get back to the point from which it
started, it will cross conductor B and induce an electromotive force into conductor A.

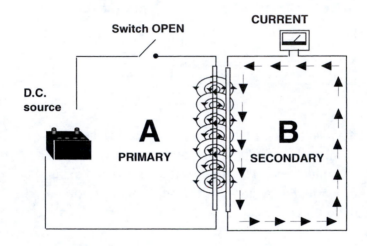

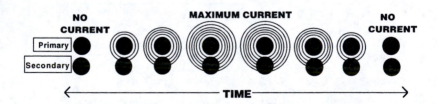

Self-induction is the effect produced by a coil which opposes any change of current in a coil.

The sketch shows only one set of magnetic lines which are expanding across the adjoining loop of wire in the coil and inducing a voltage which opposes the flow from the battery. The result of this opposition is the battery builds up current in the coil slowly. Self-induction is opposing an increase in current by generating a back pressure.

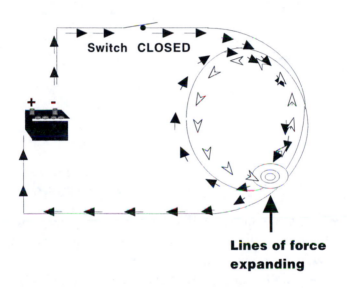

**Lines of force
expanding**

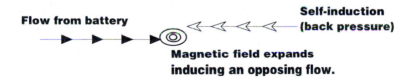

Flow from battery

**Self-induction
(back pressure)**

**Magnetic field expands
inducing an opposing flow.**

When the switch is opened the lines of force will contract back across the loop in the coil and induce a pressure in the *opposite* direction so that now instead of opposing the battery, it is assisting the battery. This action tends to keep a current flowing in the circuit and usually causes a spark at the switch when it is opened.

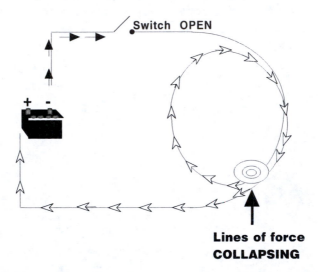

Switch OPEN

+ -

Lines of force
COLLAPSING

As you can see, self-induction always tends to *oppose a change* of current in the circuit by trying to hold the current back when the switch is closed, and by trying to keep current flowing when the switch is opened.

Perhaps you've heard of a person getting a kick from a coil even though the switch was open. When the switch is opened, the field collapses very rapidly, causing a high voltage to be induced. Since the voltage cannot force a current back to the battery after the circuit has been broken, the path will be through the body of a person to ground.

Every complete circuit has some inductance. Even the simple complete loop circuit or a single-turn coil has inductance. The coil having many turns has a higher inductance than a coil having fewer turns.

Frequency plays a role in induction also. If the frequency of AC is low, the current will have more time to reach maximum before the polarity is reversed. The higher the frequency, the lower the current through the circuit. Frequency affects the opposition to current flow as well as circuit induction.

THE VOLT-AMP

Pure inductance uses no power. The electrical energy that is taken is returned directly to the circuit as electrical energy, whereas the electrical energy taken in a *resistance is converted into heat and cannot be returned* to the circuit. *Inductance uses electrical energy to create a magnetic field* and the magnetic field restores the energy to the line when it collapses. That's why in an AC circuit volts times amps = volt-amps not watts. In an AC circuit, voltage times amps has little to do with actual power consumed, since no power (watts) is consumed in a pure inductive or capacitive AC circuit. In AC, this is called volt-amps (va), *wattless* power.

VOLT-AMPS ARE STEALING ENERGY TO CREATE A MAGNETIC FIELD AND DOES NO WORK

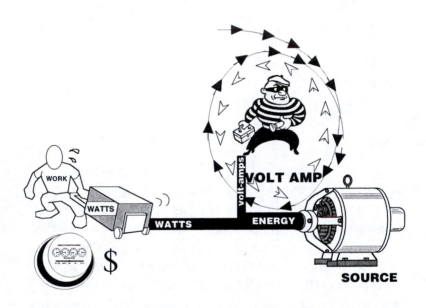

RESISTANCE IS USED UP IN HEAT (CONSUMED) DOING WORK (WATTS)

The term *"in phase"* is the portion of a cycle or period through which the current or voltage has passed since going through zero value at the beginning of the cycle or period. Phase is abbreviated: ø.

P icture two roller coaster's with cars "volts" and "amps". If the two cars start together and finish together, they are "in phase".

IN PHASE

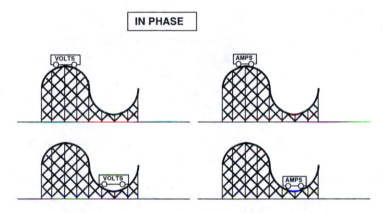

An inductive load is called a *lagging* load, by opposing the normal flow of current, the current now lags behind the voltage. This is referred to as being "out of phase". A pure resistive would be "in phase" (unity 1.0). In reality most circuits are lagging and the power factor is around 80%.

RESISTIVE LOADS
"IN PHASE"

INDUCTIVE LOADS
"OUT OF PHASE"

Even though apparent power (volt-amps) is not consumed by the circuit, it must be considered in the design of the AC generator. Generator and transformer windings must be sized to handle this current. Generators are rated according to the maximum apparent power (va) they can deliver, regardless of how much of this power is consumed by the load. Some generators may be kw rated at a certain designated power factor.

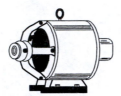

A DC generator has a power rating in kilowatts, which indicates the maximum power that the generator can constantly supply. AC generators are rated in kva since the power consumed in an AC circuit depends on the power factor. Perfect power factor is unity 1.0. In reality the power factor will always be less than 1.0.

In an AC circuit that has the inductive reactance factor, wattage (true power) is found by multiplying the voltage times amperage times the power factor.
E x I x PF = Watts.

The perfect condition is unity 1.0. This could only pertain to a system with all resistive loads.
This is not possible in the world we live today.

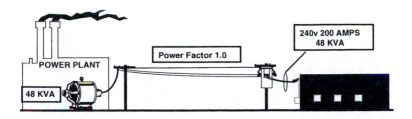

The utility company gets paid for everything used at a power factor of 1.0. 48 kva is equivalent to 48 kw.

TRUE POWER = VA X PF

TRUE POWER = E X I

240v x 200a x 1.0 = 48,000 va or 48 kva. With a unity power factor 48 kva is equal to 48 kw. At 8 cents per kilowatt hour, the utility company receives 48 kw x .08 = *$3.84 per hour*.

In the real world we cannot achieve a unity power factor with the inductive loads that we use.
The example below shows the same 48 kva, but now the *power factor is 80%*.

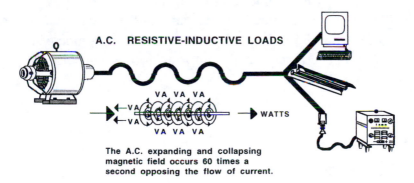

A.C. RESISTIVE-INDUCTIVE LOADS

The A.C. expanding and collapsing magnetic field occurs 60 times a second opposing the flow of current.

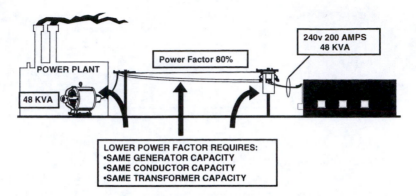

LOWER POWER FACTOR REQUIRES:
•SAME GENERATOR CAPACITY
•SAME CONDUCTOR CAPACITY
•SAME TRANSFORMER CAPACITY

The wattage (true power) is how the utility company is paid.
E x I x PF = True Power. 240v x 200a x .80 = 38, 400 watts or 38.4 kw.
38.4 kw x .08 = *$3. 07.*

Kva determines the generator size, transformer size, and conductor size. The utility company ties up the same power and conductors for $3.07 per hour as they would for $3.84 per hour.

U tility companies charge the customer for kilowatt hours consumed and can also add a penalty where the power factor is low. The utility company is tying up generator capacity and transformer capacity for low power factor loading.

SUMMARY

The Ohm's law circle actually applies to DC (direct current) pure resistance. AC (alternating current) has two components not found in a DC circuit that add opposition to the normal flow. These two components are called "inductive reactance" and "capacitive reactance".

As we have learned in theory, when an electric current moves through a wire, a magnetic field is formed around this wire. When the current in an electric circuit changes, the circuit may *oppose* the change. The property of the circuit that opposes the change is called *inductance.*

For a DC circuit, inductance affects the current flow when the circuit *is turned on or off.* When the switch is turned on, current flows through the circuit and the lines of magnetic force expand outward around the circuit conductors and the current rises from zero to its maximum value. Whenever a current flow changes, the induced magnetic field changes and opposes the change in current, whether it be an increase or decrease and inductance will *slow down* the rate at which the change occurs. When the switch is opened in a DC circuit, the current will drop very rapidly towards zero causing the magnetic field to collapse and generate a very high voltage, which not only opposes the change in current but can also cause an arc across the switch.

The big difference, is an AC circuit is *constantly* switching on and off, reversing direction 60 times a second. So circuit inductance affects AC circuits *all the time.*

In an inductive circuit when current increases, the circuit stores energy in the magnetic field. When current decreases, the circuit gives up energy from the magnetic field. In an AC circuit, the magnetic field is always changing. Every circuit has some inductance, although it may be so small that its effect is negligible.

The difference between a resistive circuit and an inductive circuit is the current flow in a resistive circuit changes *immediately* when the applied voltage is changed. In an inductive circuit current flow is *delayed* with respect to a change in applied voltage.

Only a few circuits can accomplish their purpose with *resistance* only. To ring bells, operate relays, and drive motors, the loads require *coils* for the driving force. A force that causes electrical energy to be converted into mechanical energy and mechanical energy into some form of work.

The *total* opposition to the flow of current in an *AC* circuit is called *impedance*. The impedance of an AC circuit compares with the resistance of a DC circuit. The symbol letter "Z" represents impedance which is *measured in ohms*.

Ohm's Law can be applied to an AC circuit by substituting the ohms of the *total* impedance (induction + capacitance + resistance) for the ohms resistance in DC, Ohm's Law = E = I x Z.

The definition of impede: To interfere with or slow the progress; to hinder.

Electrical inductance is like mechanical inertia. Inertia means slow to change.

When a large truck begins to move forward it is in a low gear, because the inertia of a heavy weight at rest must be overcome. As the gears are shifted the truck begins to pick up speed and the load becomes easier to move.

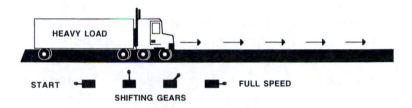

The reverse occurs when the driver wishes to stop the truck. The driver must down shift the gears to over-come the tendency of forward inertia to keep the load moving forward.

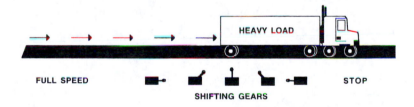

It's the same with a boat on the surface of the water. The boat begins to move at the instant a constant force is applied to it. At this instant its rate of change of speed (acceleration) is greatest. All of the applied force is used to overcome the inertia of the boat.

After awhile the speed of the boat increases (its acceleration decreases) and the applied force is used up in overcoming the friction of the water against the hull of the boat. As the speed levels off and the acceleration becomes zero, the applied force equals the opposing friction force at this speed and the inertia effect disappears.

DEFINITIONS

RESISTANCE -
Resist is to exert force in opposition.

REACTANCE -
React is to exert a reciprocal or counteracting force or influence.

INDUCTION -
Induce is to move by persuasion or influence.

CAPACITANCE -
Capacity is the potential for holding or storing.

IMPEDANCE -
Impede is to interfere with (hinder) or slow down the progress.

SYMBOL LETTERS

R = Resistance   C = Capacitance

L = Induction   Z = Impedance

The letter "X" represents reactance. Reactance is measured in ohms.

XC = Capacitive reactance   XL = Inductive reactance

The *total* opposition to the flow of current in an AC circuit is called *impedance*. The impedance of an AC circuit compares with the resistance of a DC circuit. The letter "Z" represents impedance. Impedance is measured in *ohms*.

IMPEDANCE = INDUCTION + CAPACITANCE + RESISTANCE

There are four classes of AC circuits:

I.   Resistance only

II.   Resistance and inductive reactance

III. Resistance and capacitive reactance

IV. Resistance, inductive reactance, and capacitive reactance

There is some degree of reactance in all AC circuits (in some AC circuits the reactance is so small that it is negligible).

The reactance must be combined with the ohmic resistance to determine the impedance (Z).

The impedance of an AC circuit is equal to the square root of the sum of the square of the *resistance* and the *net reactance.*

Thus: $Z = \sqrt{R^2 + X^2}$

Ohm's Law can be applied to an AC circuit by substituting the ohms of the *total impedance* (induction + capacitance + resistance) for the ohms resistance in DC, Ohm's Law: $E = I \times Z$.

ENERGY

The tendency of matter to resist change is called inertia (in nur sha). The word inert means sluggish to change and lackadaisical. The animating force that prevents inertia from taking over is called *energy*.

What does energy mean? In one sense it means power.

The word energy is derived from two Greek words, en meaning in and ergon meaning work. Energy means *in work* or the ability to do work.

Actually there are two forms of energy; *heat* and *motion*. Heat and motion in all their various forms can be converted into each other.

Heat from a gas flame causes water to steam. This is an example of heat converted to motion. And you can convert motion to heat by rubbing two sticks together. Fire is a *form* of heat as water is a *form* of motion. So heat and motion are forms of energy, *not* energy itself.

The Law of Conservation of Energy states that *energy is neither created nor destroyed.* The total amount of energy in the universe always remains the same.

Mechanical energy can be converted to electrical energy or electrical energy can be converted to mechanical energy. No matter how *energy* is changed, or what form it takes, the total amount of energy in the universe always remains the same.

Energy is motion (kinetic) or it can be stored (potential). Anything held in your hand has potential energy. When the item is dropped from your hand it creates energy with motion as it falls.

A ball held in your hand has potential energy and when thrown through the air the energy is converted to kinetic energy, as well as heat. The motion of the moving ball makes the air molecules around it start moving. The air becomes hotter and the ball itself becomes warmer. With all these forms of converted energy - the mechanical energy of moving air molecules and the total heat gain of the ball and the air around it - when added together will equal the original amount of potential energy when the ball was held in your hand.

The six major forms of energy are chemical, mechanical, electrical, light, heat, and nuclear.

Chemical energy is food and fuels. Energy can be converted into any other form of energy.

The pages of an old book turn yellow after years of time. The pages of the book are slowly releasing chemical energy as they combine with the oxygen of the air surrounding the book. The pages actually become hotter than the surrounding air and the yellowing of the pages is a result of this slow burning from chemical energy.

C hemical energy is released when complex bonds that hold matter together are broken. When atoms bond into molecules it is a potential source of energy. There is no direct way to measure it. It is possible to determine how much heat is given off when a molecule breaks during a chemical reaction.

For a long time humans had been running around the earth with human muscle power as the only source of energy.

The discovery of fire helped us warm ourselves, but the harnessing of mechanical energy led to the greatest expansion in human productive power.

The most significant change came, however, with the development of steam engines and the discovery of electricity. Now almost every corner of the world can be populated.

Electricity is one of the most important forms of energy and it cannot even be seen, heard, or smelt. The quiet giant!

A generator does not create energy. It converts mechanical energy into electrical energy. To get more electrical energy, more mechanical energy must be supplied.

Sources of energy we use are:

PETROLEUM (oil) provides almost half of the energy used in the world.

COAL developed from the remains of plants that died 400 million years ago. It provides about 30% of the world's energy.

NATURAL GAS provides about 20% of the world's energy.

Others are *WATER POWER* which costs nothing, causes no pollution, and cannot be used up. However, expensive dams or other structures are needed as the water must fall from a higher place to a lower place.

WOOD, that was once the main fuel, is still a large source of energy.

NUCLEAR ENERGY from fission is also used to generate electricity. A huge amount of energy can be obtained from a small amount of fuel.

SOLAR ENERGY is used in small applications. It is clean and unlimited, but it needs a large area of land for the collectors and is interrupted by darkness and bad weather.

Coal is abundant enough to last 300 or 400 more years. It now provides about a third of the world's energy and about a fifth of the energy used in the United States.

The amount of fossil fuels (oil, natural gas, and coal) consumed in the United States has nearly doubled every 20 years since 1900. Between 1960 and 1980, the U.S. population increased by 25%. Total energy demand, however, rose by 80%. More than three times as much as the population.

As Nations shift from agricultural to industrial economies, vast increases in energy consumption occur for powering industry and for mechanizing and fertilizing farms.

A major trend in energy demand has been a dramatic increase in the use of electricity. In 1980 the U.S. used 380 times more electricity than it did in 1900.

Usage will continue to increase because of the rising living standards. Air conditioners, dishwashers, and clothes dryers, which were luxury items in 1950, are standard appliances today.

Heated garages, escalators, temperature-controlled sports arenas, and civic centers are examples of increased energy used by the commercial sector.

Electrical energy, in vast quantities, is required for the existence of a major city to provide light, heat, transportation, and power.

Energy-intensive devices such as robots, computers, and automated systems began to replace human workers because labor costs rose faster than energy and equipment costs.

Experts forecast that fossil fuels will still meet more than 75% of the world energy needs by the year 2000. By the year 2000, hydropower is expected to decline and nuclear power to increase about 11% of the U.S. needs.

Environmental problems, particularly air polution, will restrict the use of coal. Coal-burning plants release sulphur dioxide, which combines with moisture in the air to produce acid rain. This destroys forests and lakes, and damages buildings.

Natural gas, which supplies about 25% of the U.S. energy needs, is on the decline since consumption has exceeded discovery of new sources. Experts predict that natural gas will satisfy only about 10% of the U.S. energy needs by the year 2000. The decline of reserves is a concern since natural gas is the cleanest fossil fuel.

Although hydropower is historically an important energy source, its growth potential is limited because few suitable sites exist for new hydroelectric plants.

Experts say the world's oil supply began declining around 1990, and by the year 2050 we will have used up the two trillion barrels of crude oil accumulated since prehistoric times. The conclusion is obvious and urgent! We must develop other available energy sources fast.

Solar power is still distant as the sun is too difficult to harness. It would take 8000 square miles of collectors. Example: A 7 foot diameter collector = 100 watts.

A greater growth is seen in the future for nuclear energy. Nuclear reactions are created by uranium. It is not burned like coal or oil, but produces extreme heat as the atoms are split. It's like a large furnace but no fuel is burning.

Coal costs several hundred million dollars more than nuclear power. Unlike coal there is an abundant supply of uranium. Coal fired is a thermal plant which is called a *swing* plant. As the peak times occur a thermal plant can swing into production in a matter of a few hours, where as a nuclear plant generally takes *2 days*.

As we have learned, all matter on earth, gasses, liquids, and solids are composed of invisible building blocks called atoms.

The atoms of most elements, such as iron, are stable and unchanging over time, but the atoms of some elements - radium, uranium, and others - do change. These unstable atoms are *radioactive*.

With nuclear energy the activity occurs in the nucleus of the atoms. These are called nuclear reactions, and matter *can be gained or destroyed*. The splitting of the atom is the twisting of the arm of mother nature.

Albert Einstein's formula proved that one gram of mass can be converted into a torrential amount of energy. To do this, the activity of the atoms has to occur in the nucleus.

E = energy, M = mass, and C = the speed of light which is 186,000 miles per second. When you square 186,000 you can see it would only take a small amount of mass to produce a huge amount of energy.

In the year 1930 scientists used parts of atoms such as protons, neutrons, or electrons and hurled them at nucleus at tremendous speeds so they would get inside the nucleus. With these atom smashers they discovered by splitting the atom of uranium 235 they had a process called *fission* (fish-in).

The neutron splits a nucleus in two. As it splits, the nucleus throws off extra neutrons. This chain reaction continues as billions of fission reactions happen in a second. Vast amounts of energy are given off when the nucleus changes.

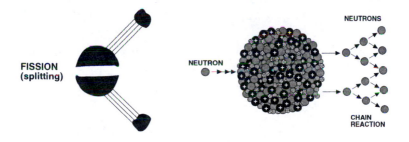

**FISSION
(splitting)**

NEUTRON

NEUTRONS

CHAIN
REACTION

Isotope is an atom that has the same number of protons but a different number of neutrons than the usual atoms of an element. Although the charge remains the same, its weight varies depending on the number of neutrons in its nucleus.

Uranium 235, the 235 refers to the isotopes weight, which is the total sum of its protons and neutrons.

In the nuclear power plant the uranium 235 pellets produce extreme heat in their tubes thus heating the water jackets to produce steam to turn the turbine generators. The danger of fission is the radioactive material called radiation.

A much safer nuclear energy is called *fusion* (fuse-in). Fusion is the joining of hydrogen atoms which causes heat hotter than the sun. Extreme heat is needed to start a fusion reaction. Most likely, lasers will be used for this source of heat. Fusion reactors are not in use yet and are still being experimented with as we have no materials to contain this extreme heat. Fusion emits no radioactive waste and is fueled by the safe and abundant hydrogen. The hydrogen atom is the lightest atom and the most susceptible to fusion. Hydrogen is the most abundant of all atoms. It can be found in every ocean.

FUSION (joining)

Fission and fusion are the two basic kinds of nuclear reactions. Fission and fusion *never* start naturally.

It took over 40 years to learn how to tame the nucleus of the atom and manipulate it to create nuclear energy.

The first atomic bomb, in 1945, was a chain reaction that raced uncontrollably. Today in nuclear power plants it can be controlled, slowed down, and completely stopped. We have come a long way since 1945.

A nuclear reactor produces heat by splitting uranium atoms. The heat boils water into high-pressure steam that drives an electrical generator.

The nuclear reactor uses *heavy water* to transfer heat and to help control the nuclear reaction.

Heavy water is a clear, colorless liquid that looks and tastes like ordinary tap water. It occurs naturally in water in minute quantities; about one part heavy water to 7,000 parts of ordinary water.

Unlike normal water, which is composed of hydrogen and oxygen, heavy water is made up of deuterium and oxygen. The name heavy water stems from the presence of deuterium, which is a form of hydrogen that has an extra neutron in its atomic nucleus and weighs slightly more than ordinary hydrogen. As a result, heavy water is about 10% heavier than normal water. It also has different freezing and boiling points.

Heavy water is about 30 times better than ordinary water in slowing down the neutrons without absorbing them.

Roughly 340,000 tons of lake water are needed to produce one ton of heavy water.

A nuclear reactor can NOT explode like a bomb. A bomb has a much higher percentage of uranium for fission.

The disadvantage is after used fuel has been removed from a reactor by remotely controlled equipment, it is stored in water-filled pools, called fuel bats, located on site at the generating station. The water cools the fuel and serves as a radiation shield.

After six years in water storage, used fuel gives off much less heat than when it first entered the fuel bay. At this time, it can either remain in water or be transferred to large, above ground dry storage containers made of steel and concrete.

These containers provide physical containment for the waste and shielding against radiation, but they have a number of advantages, including the fact that they require less maintenance than water filled storage bays. In addition, they provide secure long term storage, should there be a delay in developing a permanent disposal site for unused nuclear fuel.

SUMMARY

All electrical energy is made by changing some other form of energy into a flow of electrons.

The electricity we use today is derived by spinning magnets called generators. These generators are driven by a mechanical force.

A hydroelectric plant uses the force of falling water to turn the fan-shaped wheel of a turbine generator. About 15% of the electricity generated comes from the flow of water. Niagara Falls and Hoover Dam are examples of using falling water for mechanical energy.

When there are no waterfalls or dams nearby, fossil fuels such as coal or oil are burned to convert water to steam, which turns the fan-shaped turbine generator. 80% of the electricity used today is generated from *steam* driven turbine generators, of which 13% is nuclear powered.

We will consume more energy in the next 5 years than was consumed by all previous society in the history of the world.

Man must find new ways to create electrical energy as the fossil fuel supply dwindles. It is estimated that by the year 2020 nuclear energy will generate the huge amounts that we depend on today, the six trillion watts each day.

100 w

If you have trouble relating to six trillion watts, a human being operates on 100 watts. The brain 20 watts and the body 80 watts. You are equivalent to a 100 watt light bulb.

OHM'S LAW

A German Georg Simon Ohm (1787-1854) developed Ohm's law.

Although he discovered one of the most fundamental laws of current electricity, he was virtually ignored for most of his life by scientists in his own country.

In 1827 Georg Simon Ohm discovered some laws relating to the strength of a current in a wire. Ohm found that electricity acts like *water* in a pipe.

Ohm discovered that the current in a circuit is directly proportional to the electric pressure and inversely to the resistance of the conductors.

Ohm's Law is one of the most important things that you will use throughout your electrical career. It is a mathematical tool which is of the greatest use in determining an *unknown* factor of voltage, current or resistance in an electrical circuit in which the other two factors are known.

It is a simple law that states the relationship between voltage, current, and resistance in a mathematical equation.

In electrical terms, voltage is represented by the letter "E" (electromotive force), current by the letter "I" (intensity), and resistance by the letter "R".

The Ohm's Law formula cannot work properly unless all values are expressed in the *correct units* of measurement:

VOLTAGE is expressed in VOLTS
CURRENT is expressed in AMPERES
RESISTANCE is expressed in OHMS

We measure electromotive force in volts, we measure electric current in amps, and we measure resistance in ohms.

Electricity has many more terms that have to do with measurement: *"VOLTS, "AMPS", '"OHMS", "WATTS",* and more.

We must first understand how the electrical system functions and then mathematical analysis can follow.

Since you cannot visually *see* the flow of electrons, current, etc. and you need to see the relationship between voltage, current, and resistance, let's do it with some terms which you are more familiar with, using *water.*

WATER	ELECTRICITY
PUMP	GENERATOR
PIPE	CONDUCTOR
PRESSURE	VOLTAGE
FLOW OF GALLONS	AMPERES
RESTRICTION	RESISTANCE

(E) VOLT: The practical unit of voltage; the pressure required to force one ampere through a resistance of one ohm. To make electrons flow in a conductor, an electrical pressure must be applied and this is called electromotive force (EMF) or voltage.

(I) AMPERE: The practical unit of electric current flow; the electric current that will flow through one ohm under a pressure of one volt.

(Ω) OHM: The practical unit of electrical resistance; the resistance through which one volt will force one ampere.

(R) RESISTANCE: The opposition which a device or material offers to the flow of current; the opposition which results in the production of heat in the material carrying the current. Resistance is measured in ohms. All resistances have two dimensions: Cross-sectional area and length.

(W) POWER: The rate at which electrical energy is delivered and consumed. Power is measured in watts. A motor produces mechanical power measured in horsepower. A heater produces heat (thermal) power. A light bulb produces both heat and light power (usually measured in candlepower).

Electrical power is equal to voltage times the amperage.
$W = E \times I$

O hm's Law states: In a DC circuit, the current is directly proportional to the voltage and inversely proportional to the resistance. In other words, the water flowing in a pipe (amperage) will be increased if the water pressure (voltage) is increased. And, if the restriction (resistance) in the pipe is *less,* the water flow (amperage) will be *more*.

The *generator* is like a *water pump,* the prime mover.

The *conductor* is like the *water pipe*, the larger the conductor, the less the resistance and the more flow.

The *voltage* is like the *water pressure,* the force pushing.

The *amperes* are like the *flow of water*, an amount of current flowing is like the gallons per minute in water.

The *resistance* is like the *restriction* in the water pipe. A reduction in the water pipe size would cause opposition to the amount of gallons per minute, as would the resistor in an electrical circuit. It limits the flow of current.

Watts (power) is expressing the *rate of work* involved; the power required. With water it requires more work to pump water up to a water tower than it would to pump water at ground level. Wattage is the rate at which the electrical energy is changed into another form of energy, such as light or heat. The faster a lamp changes electrical energy, the brighter it will be.

Horsepower (hp) is the unit of measurement for mechanical power which is equal to 33,000 foot-pounds per minute. One horsepower is developed when the product of the distance and pounds equals 33,000 and this is done in one minute. In electrical terms, one horsepower = *746 watts.* One horsepower is developed if 33,000 pounds are lifted one foot in one minute. This represents the *work* done by the *output* of a motor.

To solve an unknown you will need to know two knowns.

$$E = I \times R \qquad I = E/R \qquad R = E/I$$

Put your finger on the one you want to solve and the other two knowns will show you how to solve it.

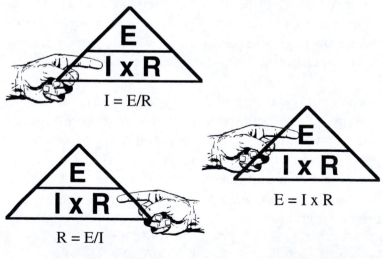

I = E/R

R = E/I

E = I x R

Get into the habit of always sketching out an Ohm's Law circuit *before* you begin trying to solve it.

$I = E/R$ One volt will force one amp through a conductor having a resistance of one ohm.

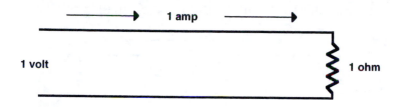

1 amp

1 volt

1 ohm

$I = E/R$ If the voltage is increased to 2 volts, the current will be 2 amps through one ohm of resistance.

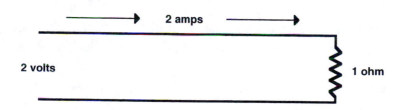

2 amps

2 volts

1 ohm

$\boxed{\text{I = E/R}}$ If the voltage is increased to 10 volts, the current will be 10 amps through one ohm of resistance.

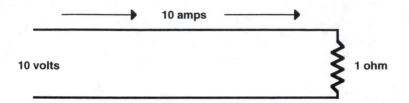

10 amps

10 volts

1 ohm

$\boxed{\text{I = E/R}}$ If the resistance is *reduced* to 1/2 ohm, the current would double to 20 amps, if the voltage remained at 10 volts.

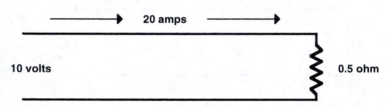

20 amps

10 volts

0.5 ohm

Directly proportional means that one factor will be *increased* in proportion to an *increase* in another factor. Example: The current *increased* to 2 amps as the voltage *increased* to 2 volts, the resistance remained the same, one ohm.

Inversely proportional means that one factor will be *increased* in proportion to a *decrease* in another factor or vice versa.

Example: The current will *increase* in proportion to a *decrease* in resistance. The current doubled to 20 amps with a decrease in resistance to 0.5 ohm.

Doubling the cross-sectional area of a conductor will reduce the resistance of the conductor by one-half.

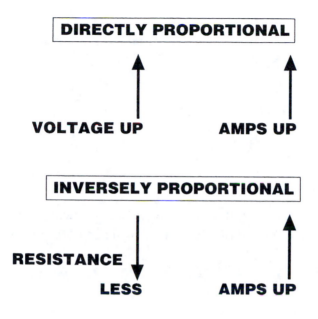

DIRECTLY PROPORTIONAL

VOLTAGE UP　　　**AMPS UP**

INVERSELY PROPORTIONAL

RESISTANCE LESS　　　**AMPS UP**

The Ohm's Law circle is divided into four parts: I = amperes; R = resistance in ohms; E = voltage; W = watts (power).

You must be able to use the circle to solve the unknown. Let's divide the circle into four parts and work examples using all the formulas.

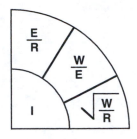

TO FIND AMPERES

$$\boxed{I = E/R}$$

Example: What is the current in amperes flowing in a circuit that has a voltage of 120 and a resistance of 10 ohms?

Solution: $I = E/R = 120v/10$ ohms = *12 amperes.*

$$\boxed{I = W/E}$$

Example: What is the current in amperes flowing in a circuit that has a 1440 watt load and a voltage of 120?

Solution: $I = W/E = 1440w/120v =$ *12 amperes.*

$$\boxed{I = \sqrt{W/R}}$$

Example: What is the current in amperes flowing in a circuit that has a 1440 watt load and a resistance of 10 ohms?

Solution: $I = \sqrt{W/R} = 1440w/10$ ohms $= 144 \quad \sqrt{144} =$ *12 amperes.*

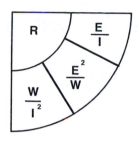

TO FIND RESISTANCE

$R = W/I^2$

Example: What is the resistance in ohms for a circuit that has a load of 1440 watts and a current in amperes of 12?

Solution: $R = W/I^2 = 1440w/144a$ (12a x 12a) = *10 ohms*.

$R = E^2/W$

Example: What is the resistance in ohms for a circuit that has a voltage of 120 and a load of 1440 watts?

Solution: $R = E^2/W = 120v$ x $120v = 14400/1440w = $ *10 ohms*.

$R = E/I$

Example: What is the resistance in ohms for a circuit that has a voltage of 120 and a current of 12 amps?

Solution: $R = E/I = 120v/12a = $ *10 ohms*.

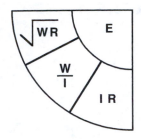

TO FIND VOLTAGE

$$E = \sqrt{W \times R}$$

Example: What is the voltage of a circuit that has a load of 1440 watts and a resistance of 10 ohms?

Solution: $E = \sqrt{W \times R} = 1440w \times 10\,ohms = 14400 \sqrt{14400}$ = *120 volts.*

$$E = W/I$$

Example: What is the voltage of a circuit that has a load of 1440 watts with 12 amps of current?

Solution: $E = W/I = 1440w/12a = $ *120 volts.*

$$E = I \times R$$

Example: What is the voltage of a circuit that has 12 amps flowing with a resistance of 10 ohms?

Solution: $E = I \times R = 12a \times 10\,ohms = $ *120 volts.*

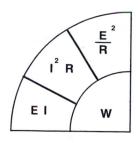

TO FIND WATTS

$$\boxed{W = E \times I}$$

Example: What is the wattage of a circuit that has a voltage of 120 with 12 amps of current flowing?

Solution: $W = E \times I = 120v \times 12a = 1440\ watts.$

$$\boxed{W = I^2R}$$

Example: What is the wattage of a circuit that has a current flowing of 12 amps and a resistance of 10 ohms?

Solution: $W = I^2R = 12a \times 12a = 144 \times 10\ ohms = 1440\ watts.$

$$\boxed{W = E^2/R}$$

Example: What is the wattage of a circuit that has a voltage of 120 and a resistance of 10 ohms?

Solution: $W = E^2/R = 120v \times 120v = 14400/10\ ohms = 1440\ watts.$

FIXED RESISTANCE

Examples of Ohm's Law applied to solve everyday situations:

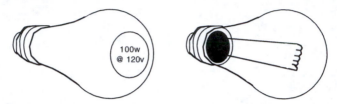

A 100 watt @ 120v light bulb is 100 watts if you apply exactly 120 volts. If the voltage is less than 120 volts, the watts will be less than 100. If the voltage is more than 120, the watts will be more than 100. The key to the first step is to find the *fixed resistance* of the bulb. This bulb was built with a fixed resistance so when you apply exactly 120 volts you will have exactly 100 watts. If the voltage is higher or lower, the wattage and current will also be higher or lower. The fixed resistance will remain the same.

Example: If you purchased a 100w @ 120v light bulb and installed it in a table lamp which had a voltage of only 115v, what is the wattage of the light bulb?

Solution: First step find the *fixed resistance*, $R = E^2/W = $ 120v x 120v/100w = 144Ω. $W = E^2/R = $ 115v x 115v/144Ω = *91.84 watts*.

Example: The output air of an electric heat unit is warm but not really hot. Check the name plate for kw and volts, now check the actual voltage supplying the unit. If the nameplate on the unit reads 10 kw @ 240v and the source to the unit is only *208 volts,* what is the actual kw output?

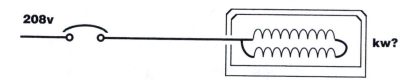

208v   **kw?**

Solution: First step find the *fixed resistance,* $R = E^2/W =$ 240v x 240v/10,000w = 5.76Ω. $W = E^2/R =$ 208v x 208v/ 5.76Ω = 7511w. 7511w/1000 = *7.511 kw.*

TUNGSTEN METAL FILAMENT   A higher resistance wire is used for lighting and heat. A light bulb filament contains tungsten metal which offers a high resistance to the flow of current thus producing light.

An electric heating element (coil) contains a higher resistance nichrome wire which offers opposition to the flow of current thus producing heat (wattage).

SERIES CIRCUIT

An electric circuit is a complete path through which electrons can flow from the negative terminal of the voltage source, through the connecting wires, through the load or loads, and back to the positive terminal of the voltage source. A complete circuit is made up of a voltage source, connecting wires, and the effective load. If the current can't get back to the source of supply, it will never leave.

If the circuit is arranged so that the electrons have only *one possible path*, the circuit is called a *series circuit.*

In a series circuit all devices are connected end to end, in a closed path, and the *same* amount of current flows through each device.

If two or more resistances are connected in such a way that they carry the *same* current, they are in *series.*

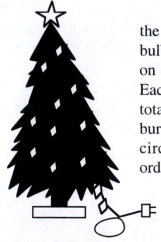

The series circuit was used in the old-style Christmas lights. Each bulb was rated at 15 volts when used on a 120 volt circuit of eight lights. Each bulb received one-eighth of the total 120 volts or 15 volts. If one bulb burns out, they all go out. Series circuit wiring is impractical for ordinary purposes.

Example: What is the current flowing in this series circuit?

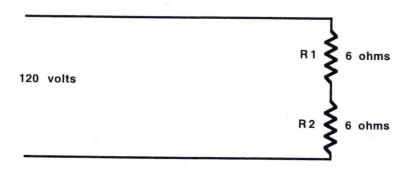

Solution: The first step is to find the resistance in the series circuit. The *total* resistance is equal to the *sum* of the individual resistances.

Total resistance in series : R total = R1 + R2
6 ohms + 6 ohms = 12 ohms

Now apply the Ohm's Law formula for current: I = E/R
E = 120 volts   120v/12Ω = *10 amps current flowing.*

Since there is but *one* path for current to flow in a series circuit, the same current must flow through each part of the circuit. To determine the current throughout a series circuit, only the current through *one* of the loads need be known.

Example: With 6 amps of current flowing in this series circuit, what is the applied voltage?

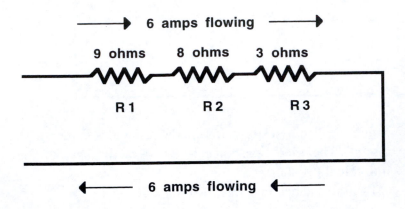

Solution: The first step is to find the total resistance. In a series circuit the formula for total resistance is:
  R1 + R2 + R3 + ...   9Ω + 8Ω + 3Ω = 20 Ω R total.

  Now apply the Ohm's Law formula: E = I x R
The current is 6 amps x 20 ohms = *120 volts.*
120 is the applied voltage to this series circuit.

In the basic series circuit with only *one* resistor, the voltage drop across the resistor is the total voltage across the circuit and is equal to the applied voltage. In a series circuit with *more* than one resistor, the total voltage drop is also equal to the applied voltage, but consists of the sum of two or more individual voltage drops. In any series circuit the *sum* of the resistor voltage drops must equal the source voltage.

This can be proven with the three resistor series circuit shown above. Voltage drop $= I \times R$. 6 amps x 9 $\Omega = 54$ volts dropped. It takes 54 volts to force 6 amps through a 9 ohm resistor. 6 amps x 8 $\Omega = 48$ volts dropped. 6 amps x 3 $\Omega = 18$ volts dropped. Total voltage drop $= 54v + 48v + 18v = 120$ volts. But remember, even though this voltage is used to push the load through the resistors and it drops from 120 volts to zero volts, the voltage is replaced every *1/60 th of a second* in a 60 cycle system.

Example: With 11.5 amps flowing in this series circuit, what is the total resistance?

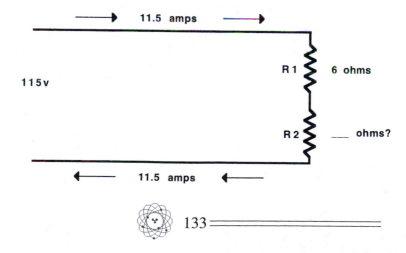

Solution: Apply Ohm's Law formula $R = E/I = 115v/11.5a$ = 10 ohms   R total = *10 ohms.*

Example: What is the resistance of R2?

Solution: 10 ohms R total - 6 ohms R1 = *4 ohms R2.*

Example: What is the voltage at R1?

Solution: Apply Ohm's Law formula $E = I \times R = 11.5a \times 6\Omega$ = *69 volts.*

Example: What is the voltage at R2?

Solution: Apply Ohm's Law formula $E = I \times R = 11.5a \times 4\Omega$ = *46 volts.*

•Checkpoint: 69 volts + 46 volts = 115 volts (the applied voltage).

 Current flows the same in series. It takes 115 volts to push 11.5 amps through 10 ohms of resistance.

Each of the resistors in a series circuit consumes power which is dissipated in the form of heat. The total power (watts) must be equal in amount to the power consumed by the circuit resistors. In a series circuit the total power is equal to the *sum* of the wattage dissipated by the individual resistances.

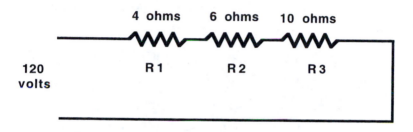

$W = E^2/R$ 120v x 120v $= 14400$ R = total resistance 4Ω + 6Ω + 10Ω = 20Ω R total   14400/20Ω = *720 watts.*

What is the current flowing in this series circuit? $I = E/R = 120v/20Ω = 6$ *amperes.* Current flows the same through a series circuit.

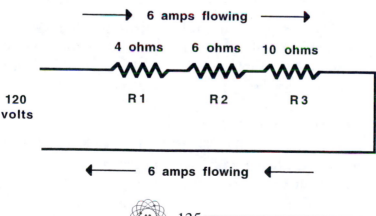

The wattage consumed at R1 is $W = I^2R = 6$ amps x 6 amps x 4Ω = *144 watts.*

The wattage consumed at R2 = 6 amps x 6 amps x 6Ω = *216 watts.*

The wattage consumed at R3 = 6 amps x 6 amps x 10Ω = *360 watts.*

Total consumed wattage = 144w + 216w + 360w = *720 watts.*

A series circuit has loads connected so there is only *one path* for current to flow.

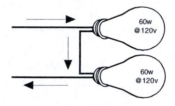

When two light bulbs are connected in series, current flows from the source into one bulb, then flows into the second bulb, and then flows back to the source of power completing the circuit. If either bulb burns out, the entire circuit is open.

The sum of the voltage drop across both bulbs would equal the source voltage. Example; if you had two identical light bulbs of 60 watts and a source voltage of 120 volts, the voltage drop across each bulb would be 60 volts.

The resistance of each bulb would be 120v x 120v/ 60w = 240 ohms. The resistance in series adds together collectively, 240Ω + 240Ω = 480Ω total resistance in the series circuit.

The current flowing in this circuit would be 120v/480Ω = .25 amps. The current flow would be the same through the entire circuit.

The wattage would be 120v x .25 amps = 30 watts. The light bulbs will be dim.

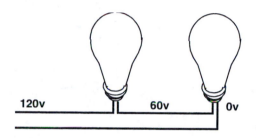

Wattage is work being done, which is why we use electricity. When a resistance (opposition) of 240Ω is placed in the circuit, the entire 120 volts is used to force the .25 amps through the resistance. On a 60 cycle AC system the 120 volts is being replaced by the source generator 60 times a second.

This circuit would look like this on a drawing:

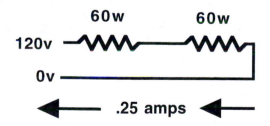

The next example shows a 60 watt light bulb and a 75 watt light bulb connected in series.

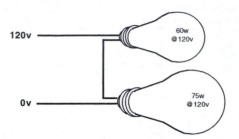

T he resistance of the 60 watt bulb would be 120v x 120v/60w = 240 ohms. The resistance of the 75 watt bulb would be 120v x 120v/75w = 192 ohms. The total resistance in series would be 240Ω + 192 Ω = 432 ohms.

Current flow is the same through a series circuit. The current would be 120v/432Ω = .28 amps.

The voltage drop across the 60w bulb would be 240Ω x .28a = 67 volts. The voltage drop across the 75w bulb would be 192Ω x .28a = 53 volts. The total voltage drop in the circuit would be 67v + 53v = 120 volts.

The total wattage would be 120v x .28 amps = 34 watts. Again very dim.

S hown below are two different sizes of water pipes. Let's say the first pipe is 3/4" diameter and the second pipe is a larger 1" diameter. The water is being forced through the first pipe at 120 pounds of pressure and through the second pipe at 53 pounds pressure due to the drop in pressure. Even though the larger pipe has a smaller pressure, the water flow through the pipe is the same as the first pipe. Always remember in a *series* circuit, current flow is the same through the entire circuit.

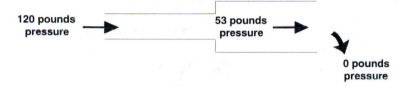

120 pounds
pressure

53 pounds
pressure

0 pounds
pressure

You can see now why series circuits are not used for lighting circuits in a building. If one bulb burns out, all the lights go out as the entire circuit is *open*.

Series circuits are used in control circuits with stop buttons or overload interlocks.

SUMMARY: SERIES CIRCUIT

•The same current flows through each part of a series circuit.
•The total resistance is equal to the sum of individual resistances.
•The total voltage across a series circuit is equal to the sum of individual voltage drops.
•The voltage drop across a resistor is proportional to the size of the resistor.
•The total power dissipated is equal to the sum of the individual power dissipations.

PARALLEL CIRCUIT

A parallel circuit is a circuit having *more than one path* for current to flow from a common voltage source.

SERIES CIRCUIT HAS ONLY ONE PATH

PARALLEL CIRCUIT HAS MORE THAN ONE PATH

T he old-style Christmas lights were connected in series. The current had only one path to flow as the lights were connected end to end in series. If one light bulb burnt out the entire circuit was opened.

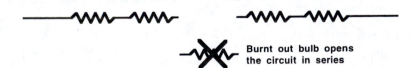

Burnt out bulb opens the circuit in series

In parallel the lights are connected *side-by-side* instead of end-to-end so that there exists more than one path through which current can flow.

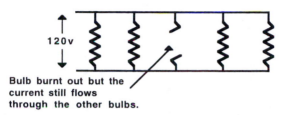

120v

Bulb burnt out but the current still flows through the other bulbs.

In a series circuit a *portion* of the source voltage is dropped across each series load and the sum of these individual voltage drops is equal to the source voltage.

When resistances in parallel are connected across a voltage source, the *voltage* across each of the resistors will always be the same. However, the *entire* source voltage is dropped across each load. This is the reason that all parallel loads are connected directly across the voltage source.

When a series circuit becomes short-circuited, the resistance of the other loads in series keeps the circuit resistance from dropping to zero. Parallel circuits develop *larger* damaging short circuit currents because each parallel load is connected directly across the source voltage. If any one of the parallel loads becomes shorted, it drops the resistance between the load and the source to practically zero.

The *current* through each resistor will vary depending on the size of each individual resistor.

In a parallel circuit, loads having *low resistance* draw *more* current than loads having high resistance.

Parallel circuits have two types of current flow; *total current flow* and the current flow through each *individual* load.

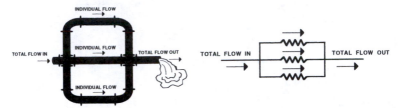

To calculate the current flowing in a parallel circuit start by calculating the current flow in *one* load.

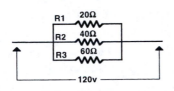

$I = E/R$

The current flowing through R1 = 120v/20Ω = 6 amperes.
The current flowing through R2 = 120v/40Ω = 3 amperes.
The current flowing through R3 = 120v/60Ω = 2 amperes.

The *total current* in parallel is equal to the *sum* of all the individual load currents.

I1 + I2 + I3 = I total   6 amperes + 3 amperes + 2 amperes = 11 amperes.

The total current flow in is 11 amperes and the total current flow out is 11 amperes. The 11 amperes is divided up between three individual loads as it flows through the parallel circuit.

In a *series* circuit the total resistance was solved by adding all of the individual resistances together. The more resistances there are, the more the total resistance would be.

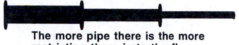

The more pipe there is the more restriction there is to the flow. The water has only one path to follow.

In a parallel circuit the *total* resistance is *less* than the size of the smallest load. Everytime you *add* another load to the circuit the *total* resistance will be *less*.

Everytime you add another amount the total is more! Let's put these words into a *picture* and *see* how it is easy to understand.

Each time a pipe (load) is *added* the total restriction (resistance) is *less*. By adding another pipe you create another path for the flow. With the addition of another pipe you have more flow (amperage). The only way you can have more flow (amperage) is to have less restriction (resistance).

EQUAL PARALLEL RESISTANCES

The simplest calculation for total resistance in a parallel circuit is when all of the loads are *equal* in resistance.

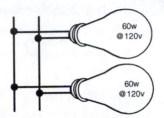

A parallel circuit has multiple loads connected across the source voltage. Each load in a parallel circuit works independently from the other loads. Current does not have to flow through one load to reach another load as they did in a series circuit. With each load having its own separate path for current flow from the voltage source, the amount of current that flows through each load is decided by the resistance of that load alone and the source voltage. The total amount of current in a parallel circuit is equal to the sum of the currents flowing through each load.

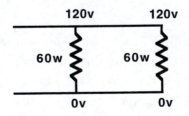

A parallel circuit would look like this in a drawing. If one bulb would burn out, the circuit would not become open as the other bulb has its own independent voltage source.

The total resistance of a parallel circuit is less than the resistance of the smallest load resistance.

The resistance of each 60 watt bulb is 120v x 120v/60w = 240 ohms. In a series circuit we added the resistances together because the current has to flow through both loads as it only had one path. The same as a single water pipe. It has only one path to follow.

The parallel circuit would have two paths for the current to flow. With two water pipes there is less restriction of the flow of water than with only one pipe. Now you have another path to flow.

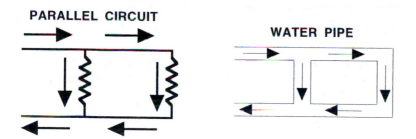

In parallel the total resistance in the circuit would be less than any one resistance. In this circuit we have two 60 watt bulbs each having a resistance of 240 ohms.

The simplest calculation for total resistance in a parallel circuit is when all of the loads are *equal* in resistance.

When two pipes of equal size are combined together (paralleled), the two together have twice the cross-sectional area. They could, therefore, be replaced with one pipe and it would double the cross-sectional area.

The same is true with conductors. By paralleling two conductors together the resistance of one conductor is cut in half because the cross-sectional area has doubled.

By using the formula below it is very easy to solve the total resistance in parallel for loads of equal resistance. $240\Omega/2 = 120$ ohms total resistance for the two 60 watt bulbs in parallel.

$$\textbf{TOTAL RESISTANCE IN PARALLEL FOR EQUAL RESISTANCES} = \frac{\textbf{RESISTANCE OF ONE}}{\textbf{NUMBER OF RESISTANCES}}$$

The total current flow would be $120v/120\Omega = 1$ amp. The total current can be solved by adding together the sum of the current flowing in each load.

The current flow each load is $120v/240\Omega = .5$ amp. Adding together $.5 + .5 = 1$ amp.

As shown below the 1 amp current flow has two paths to divide. These two paths both have 240 ohms of resistance so the current flow divides equally with .5 amp flowing through each load and joining together again for a total flow in the circuit of 1 amp.

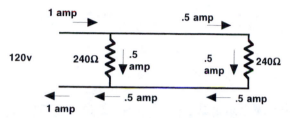

Three equal resistances in parallel would calculate:

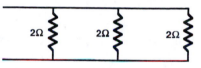

The resistance of one = 2Ω/3 the number of resistances = .66 Ω total resistance.

Every time you *add* a load to a parallel circuit the total resistance will go *DOWN*. Four equal resistances in parallel would calculate:

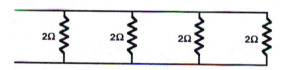

The resistance of one = 2Ω/4 the number of resistances = .5 Ω total resistance.

We have learned to relate electrical circuits to the flow of water. As shown below, everytime you add a pipe (load) the restriction (resistance) to the flow (amperes) of water (source) is less as you have added another path of flow.

 147

In practical wiring applications we are connecting our lighting circuits in parallel.

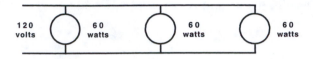

Example: What is the total current in this parallel circuit?

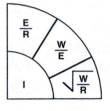

Solution: Apply Ohm's Law formula I = W/E = 60w/120v = 0.5 amp per light. I total = I1 + I2 + I3 = 0.5a + 0.5a + 0.5a = *1.5 amps total current in parallel.*

Example: What is the resistance of one light?

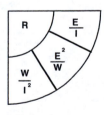

Solution: Apply Ohm's Law formula R = E²/W = 120v x 120v/60w = *240 ohms each light.*

Example: What is the total resistance of this parallel circuit?

Solution: Use the formula for *equal* resistors in parallel 240Ω/3 resistors = *80 ohms R total.*

UNEQUAL PARALLEL RESISTANCES

When a circuit contains resistors in parallel with different *unequal* values, the problem of solving the total resistance becomes more difficult.

There are different formulas for *unequal* parallel resistors you can apply, both giving the same results.
1/Rt = 1/R1 + 1/R2 + 1/R3 .....

Example: What is the total resistance of this parallel circuit?

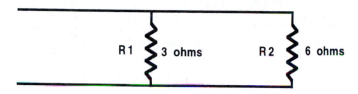

1/Rt = 1/3Ω + 1/6Ω   the common denominator is 6.
2/6 + 1/6 = 3/6   now invert 6/3 = *2 ohms total resistance.*

The other formula to find total resistance *unequal* in parallel is: $\dfrac{R1 \times R2}{R1 + R2}$

R1 = 3 ohms   R2 = 6 ohms $\dfrac{3Ω \times 6Ω}{3Ω + 6Ω} = \dfrac{18}{9} = $ *2 ohms total resistance.*

By using either formula you will get the *same* answer, 2 ohms total R, which is *less* than any one resistance.

Example: What is the total resistance for *three unequal* resistors in parallel?

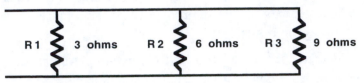

Solution: Use the formula 1/Rt = 1/R1 + 1/R2 + 1/R3. To find the common denominator multiply R1 x R2, if R3 will divide *evenly* that will be the common denominator. 3 x 6 = 18/9 = 2 (divides evenly).

$$\frac{1}{3} + \frac{1}{6} + \frac{1}{9}$$ 18 is the common denominator.

$$\overline{18} + \overline{18} + \overline{18}$$

$$\frac{6}{18} + \frac{3}{18} + \frac{2}{18} = \frac{11}{18}$$ invert to 18/11 = *1.63 ohms total resistance.*

    Now apply the other formula for *unequal parallel.* For 3 resistors the formula would be:

$$\frac{R1 \times R2}{R1 + R2} = Y \qquad \frac{Y \times R3}{Y + R3} = Rt$$

R1 = 3   R2 = 6   R3 = 9

$$\frac{3 \times 6}{3 + 6} = \frac{18}{9} = 2 \qquad Y = 2$$

$$\frac{2 \times 9}{2 + 9} = 18 = \textit{1.63 ohms total resistance.}$$

Here is another way to find the total resistance in parallel for unequal resistors using your *calculator*.

Let's use the same resistors 3 Ω, 6 Ω and 9 Ω and see if we get the same answer of 1.63 Ω R total.

Simply follow these steps:

Using your calculator, clear it so it reads $\boxed{0.}$ Make sure it does *not* read $\boxed{0.^M}$

Press $\boxed{1}$ Press $\boxed{\div}$ Press $\boxed{3}$ Press $\boxed{M+}$ Your calculator should read 0.3333333^M

Press $\boxed{1}$ Press $\boxed{\div}$ Press $\boxed{6}$ Press $\boxed{M+}$ Your calculator should read 0.1666666^M

Press $\boxed{1}$ Press $\boxed{\div}$ Press $\boxed{9}$ Press $\boxed{M+}$ Your calculator should read 0.1111111^M

Press $\boxed{1}$ Press $\boxed{\div}$ Press $\boxed{M_C^R}$ Press $\boxed{=}$ Your calculator should read 1.6363639^M

The numbers $\boxed{3}$ $\boxed{6}$ $\boxed{9}$ are for the unequal resistors of 3 Ω, 6 Ω, and 9 Ω. By using your calculator you can calculate the total resistance for as many resistors as the circuit would have by simply following this format. The *final step* is always :

Press $\boxed{1}$ Press $\boxed{\div}$ Press $\boxed{M_C^R}$ Press $\boxed{=}$ Your calculator should read the *ANSWER.*

Practice this calculator drill a few times, it can be very helpful in determining total resistance in a parallel circuit with *unequal* resistors.

Example: Find the total resistance in this parallel circuit having *four unequal* resistors.

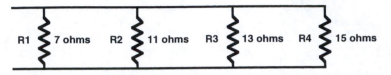

Solution: 1/Rt = 1/R1 + 1/R2 + 1/R3 + 1/R4
Find the common denominator 7 x 11 x 13 x 15 = 15,015.
The common denominator is 15,015.

$$\frac{1}{7} + \frac{1}{11} + \frac{1}{13} + \frac{1}{15}$$

$$\frac{2145}{15015} + \frac{1365}{15015} + \frac{1155}{15015} + \frac{1001}{15015} = \frac{5666}{15015}$$

invert 15015/5666 = *2.65 ohms total resistance*

Using the other formula:

$$\frac{R1 \times R2}{R1 + R2} = Y \qquad \frac{Y \times R3}{Y + R3} = Z \qquad \frac{Z \times R4}{Z + R4} = Rt$$

R1 = 7   R2 = 11   R3 = 13   R4 = 15

$$\frac{7 \times 11}{7 + 11} = \frac{77}{18} = 4.28 \qquad Y = 4.28 \qquad \frac{4.28 \times 13}{4.28 + 13} = \frac{55.64}{17.28} = 3.22$$

$$Z = 3.22 \qquad \frac{3.22 \times 15}{3.22 + 15} \qquad = \qquad \frac{48.3}{18.22} \qquad = \quad 2.65\Omega$$

R total = 2.65 ohms.

In a parallel circuit the total wattage is *equal* to the sum of the wattage dissipated by the individual resistances, same as in a series circuit.

Example: To find the total wattage in the parallel circuit below there are several ways to calculate the power.

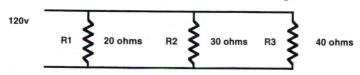

$W = E \times I$
$I = E / R$

$120v/20\Omega = 6a$
$120v/30\Omega = 4a$
$120v/40\Omega = 3a$

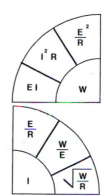

I total $= 6a + 4a + 3a = 13$ amps

Total wattage $= 120$ volts x 13 amps $= $ *1560 watts*

The wattage in each branch $= I^2R$ and then add the branches.
R1 $= 6a \times 6a \times 20\Omega = 720$ watts
R2 $= 4a \times 4a \times 30\Omega = 480$ watts
R3 $= 3a \times 3a \times 40\Omega = \underline{360 \text{ watts}}$
1560 watts

SUMMARY: PARALLEL CIRCUIT

• The total current is equal to the sum of the branch currents.
• The total resistance is less than any one individual resistance.
• The voltage across each branch is the same as the source voltage.
• The total power consumed is equal to the sum of the individual power dissipations.

SERIES - PARALLEL CIRCUIT

A *series-parallel* circuit is a combination of both a series and a parallel circuit. Some parts of the circuit are connected in series and some parts are connected in parallel.

In order to analyze the circuit you must be able to *recognize* which parts are connected in series and which parts are connected in parallel. With some circuits it will be easy to recognize how it's connected. With other circuits it will be more difficult and you will have to *redraw* the circuit and put it in the *simplest form*.

Here are some reminders that we have read.

SERIES CIRCUIT HAS ONLY ONE PATH

PARALLEL CIRCUIT HAS MORE THAN ONE PATH

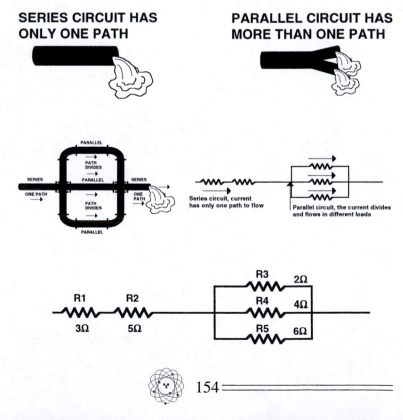

To find the *total resistance* of this series-parallel circuit start at the end of the circuit and calculate the resistance of R3, R4, and R5 that are connected in parallel.

$$\frac{2\Omega \times 4\Omega}{2\Omega + 4\Omega} = \frac{8}{6} = 1.33 \quad \frac{1.33 \times 6\Omega}{1.33 + 6\Omega} = \frac{7.98}{7.33} = 1.088 \text{ or } 1.09$$
total resistance

Now R3, R4, and R5 have been reduced to their simplest form of *one resistor* with a resistance of 1.09Ω.

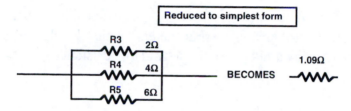

R1 and R2 are connected in series so they would *add* together for a total resistance in series of $3\Omega + 5\Omega = 8\Omega$ *total series resistance.*

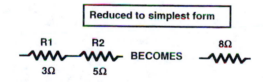

Reduced to its simplest form the circuit now looks like this:

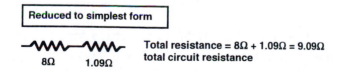

155

Example: Find the total resistance in this series-parallel circuit.

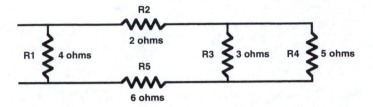

Solution: When trying to solve total resistance in a series-parallel circuit always start at the *end* of the sketch. R4 is connected in parallel with R3 so find the resistance of these two.

$$\frac{R3 \times R4}{R3 + R4} = \frac{3 \times 5}{3 + 5} = \frac{15}{8} = 1.875 \text{ ohms}$$
(combined resistance of R3 and R4)

Now the circuit looks like this:

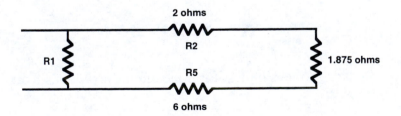

1.875 ohms is in series with R2 and R5. Series resistance adds together, $1.875\Omega + 2\Omega + 6\Omega = 9.875\Omega$.

The circuit now looks like this:

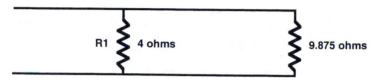

R1 4 ohms   9.875 ohms

9.875 ohms is in parallel with R1. $\dfrac{9.875 \times 4}{9.875 + 4} = \dfrac{39.5}{13.875} =$

2.84 ohms total resistance.

In practical wiring applications we use series-parallel connections for heating elements in cooking units to obtain different heat selections.

COOKING ELEMENTS

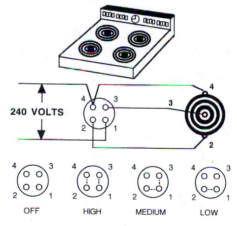

240 VOLTS

OFF   HIGH   MEDIUM   LOW

The heating element is connected to a special control switch 240/120v which allows different wattages to be obtained by different switch positions.

The element would look like this in a sketch:

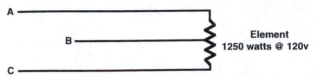

Example: Find the fixed resistance of the cooking element.

Solution: Use Ohm's Law formula $R = E^2/W = 120v \times 120v/ 1250w = $ *11.52 ohms fixed resistance.*

For *high heat* 240volts would be applied through wires A and C through the special control switch.

High heat wattage : $W = E^2/R = 240v \times 240v/11.52\Omega = $ *5000 watts.*

For *medium heat* 120 volts would be applied through wires A and B. Now you are using half of the resistance, 5.76 ohms, by tapping from wire B.

Medium heat wattage: $W = E^2/R = 120v \times 120v/5.76\Omega = $ *2500 watts.*

For *low heat* 120 volts would be applied through wires A and C. Now you are using the full resistance 11.52 ohms at the low voltage 120.

Low heat wattage: $W = E^2/R = 120v \times 120v/11.52\Omega = $ *1250 watts.*

THE PATH OF CURRENT FLOW

Current will flow in the path of the least resistance. The following is an example of tracing the path through a complete circuit from the source and back to the source.

T he circuit shows four switches (sw1, sw2, sw3, sw4) and five loads (10 ohm, 10 ohm, 5 ohm, 20 ohm, 20 ohm) connected in the circuit.

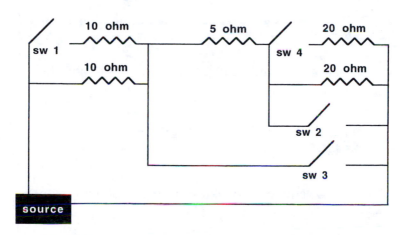

With all of the switches in the "open" position the path of current flow starts at the source and flows through one 10 ohm load. The current cannot flow through the other 10 ohm load because switch one (sw 1) is in the open position. The current continues on through the 5 ohm load as the other path (wire) goes to switch two (sw 2) and it is in the open position. Since switch four (sw 4) is also open, the current will flow through one 20 ohm load and back to the source of supply. ◄—

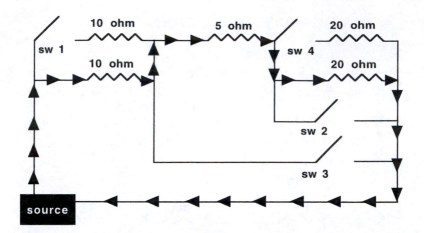

The circuit is actually a series circuit and looks like this:

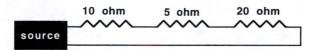

The example below shows the circuit with switch one and three (sw 1 & sw 3) closed.

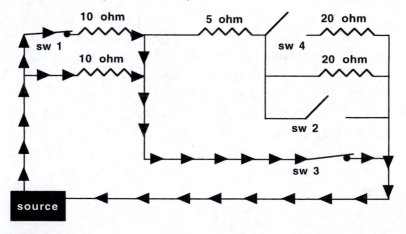

N ow the path of current flow is through *both* 10 ohm loads as "sw 1" is closed. The path will now go through "sw 3" since it is closed and is the path of *least* resistance compared to the other path, which is the wire to the 5 ohm load.

The circuit is now a parallel circuit with two equal 10 ohm loads.

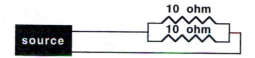

The total resistance to the flow of current is 10 ohm divided by two = 5 ohms total resistance in the circuit with switches one and three closed.

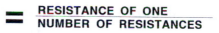

TOTAL RESISTANCE IN PARALLEL FOR EQUAL RESISTANCES = **RESISTANCE OF ONE** / **NUMBER OF RESISTANCES**

The example now shows the circuit with switch one and four (sw 1 & sw 4) closed.

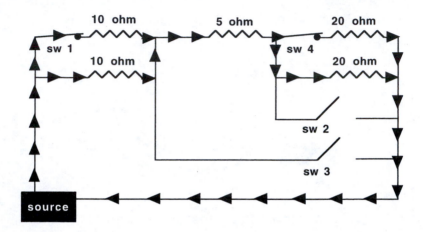

Now the path of current flow is through both 10 ohm loads and on through the five ohm load and continues through both 20 ohm loads as switch four is closed.

The circuit now is a series-parallel circuit.

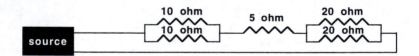

To solve the total resistance to the flow of current in this series-parallel circuit start at the *end* of the circuit which is two 20 ohm loads in parallel. The resistance of one is 20 ohms divided by 2 resistors equals 10 ohms. The circuit now looks like this:

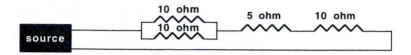

The next step in solving the total resistance in the circuit would be to combine the other two parallel loads. The resistance of one is 10 ohms divided by 2 resistors equals 5 ohms. The circuit now looks like this:

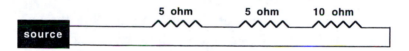

The total resistance to the flow of current with switches one and four closed is 5 ohm + 5 ohm + 10 ohm = 20 ohms total resistance.

SUMMARY - THE CIRCUIT

The electric circuit is a path in which an electric current flows. A water circuit is a path in which water flows.

Before water can flow in a pipe, some means must be used to force the water to move. A water pump is used to provide the pressure to force the water to move through the pipes. The pipes offer resistance to the flow of water. The water motor is the load. This motor may be used to drive equipment.

In an electric circuit, the current must flow through a set of wires called conductors. These wires act just the same in an electric circuit as pipes do in a water circuit.

In the water circuit all of the water which leaves the pump must return to the pump. No water is lost in the circuit. The same holds true in the electric circuit. The current which leaves one terminal of the generator, returns to the other terminal and is pumped out into the circuit again.

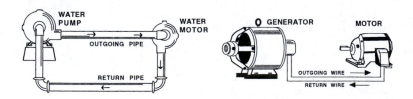

The rate at which water flows through a pipe is measured in gallons per minute, but in the electric circuit, the current flow is measured in amperes.

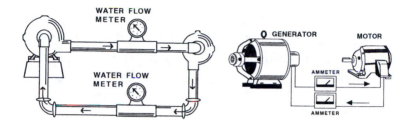

The water flow meters will both read the same, as the water on the outgoing pipe will be the same on the return pipe.

The ammeters will both read the same, as the current on the outgoing wire will be the same on the return wire.

Resistance is opposition to the flow of current in a wire. All wires have resistance. The more resistance a conductor has, the more difficult it will be for current to flow through it. Resistances are measured in ohms. There is no term or unit to measure the resistance of a water pipe.

Conductivity is the opposite of resistance. The better the conductivity the lower the resistance.

In a water circuit the valve is used to govern the flow of water. By closing the valve it will add more resistance to the flow of water.

A rheostat (adjustable resistor) will govern the flow of current in an electrical circuit by changing the resistance.

The rheostat has a movable arm which makes contact with the resistance wire. As the arm is moved to the left, the current must flow through more of this wire which adds resistance and less current will flow in the circuit.

165

he water valve does the same thing in a water circuit. Since it will be difficult for water to flow through the small opening in the valve, pressure will be lost and the amount of water flowing will be cut down.

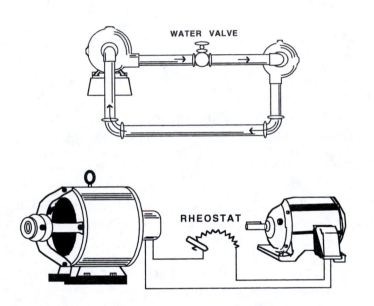

WATER VALVE

RHEOSTAT

It's easier for water to flow through a big pipe than a smaller one. It's the same with electrical conductors. The larger the wire, the less resistance it has.

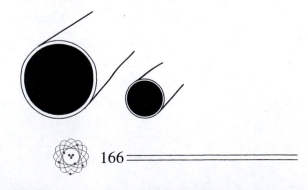

REVIEW

QUESTIONS

EXAM #1

1. The voltage per turn of the primary of a transformer is _____ the voltage per turn of the secondary.

(a) more than
(b) the same as
(c) less than
(d) none of these

2. As the power factor of a circuit is increased _____.

(a) reactive power is decreased
(b) active power is decreased
(c) reactive power is increased
(d) both active and reactive power are increased

3. The AC system is preferred to the DC system because _____.

(a) DC voltage cannot be used for domestic appliances
(b) DC motors do not have speed control
(c) AC voltages can be easily changed in magnitude
(d) high-voltage AC transmission is less efficient

4. Inductance in a circuit _____.

(a) delays the change in current
(b) prevents current from changing
(c) causes power loss
(d) causes current to lead voltage

EXAM #1

5. Low power factor in an industrial plant generally is caused by ____.

(a) excessive inductive loads
(b) excessive resistive loads
(c) insufficient induction load
(d) insufficient resistance load

6. If 18 resistances each rated 36 Ω are connected in parallel, the total resistance would be ____ Ω.

(a) 648 (b) 324 (c) 9 (d) 2

7. A negatively charged body has ____.

(a) excess of electrons   (b) excess of neutrons
(c) deficit of electrons   (d) deficit of neutrons

8. A capacitor stores ____.

(a) voltage (b) power (c) current (d) charge

9. Electrical appliances are not connected in series because ____.

(a) series circuits are complicated
(b) appliances have different current ratings
(c) the power loss is too great
(d) the voltage is the same

10. A capacitor opposes _____.

(a) both a change in voltage and current
(b) change in current
(c) change in voltage
(d) none of these

11. Permanent magnets use _____ as the magnetic material.

(a) nickel   (b) iron   (c) hardened steel   (d) soft steel

12. A component having no continuity would have _____ resistance.

(a) high   (b) low   (c) infinite   (d) none of these

13. The total opposition to the flow of alternating current is _____.

(a) resistance   **(b) impedance**
(c) induction   **(d) capacitance**

14. 3 ø currents are generally out of phase by _____ degrees.

(a) 30   (b) 60   (c) 90   (d) 120

15. _____ is the ratio of output to input.

(a) Reluctance   **(b) Cosine**
(c) Efficiency   **(d) Square root**

EXAM #1

16. The voltage produced by electromagnetic induction is controlled by _____.

(a) the number of lines of flux cut per second
(b) eddy currents
(c) the size of the magnet
(d) the number of turns

17. Which of the following has the highest dielectric strength to electrical breakdown?

(a) thermoplastic   **(b) impregnated paper**
(c) rubber   **(d) woven cloth**

18. The greatest voltage drop in a circuit will occur when the _____ the current flow through that part of the circuit.

(a) greater   (b) slower   (c) faster   (d) lower

19. _____ results in loss of electrical energy from the circuit.

(a) Resistance   **(b) Reluctance**
(c) Susceptance   **(d) Admittance**

20. Soft iron is most suitable for use in a _____.

(a) natural magnet   **(b) permanent magnet**
(c) magneto   **(d) temporary magnet**

EXAM #2

1. A length of wire has a resistance of 6 ohms. The resistance of a wire of the same material three times as long and twice the csa will be _____ ohm/s.

(a) 36   (b) 12   (c) 9   (d) 1

2. The electrons in the last orbit of an atom are called _____ electrons.

(a) bound   (b) free   (c) valence   (d) atomic

3. A substance whose molecules consist of the same kind of atoms is called _____.

(a) proton   (b) valence   (c) element   (d) compound

4. Other factors remaining the same, the effect on the current flow in the circuit would cause the current to _____ if the applied voltage was doubled.

(a) double   (b) divide by 2
(c) remain the same   (d) increase 4 times

5. In a series circuit when the voltage remains constant and the resistance increases, the current _____.

(a) increases   (b) decreases
(c) remains the same   (d) increases by the square

EXAM #2

6. Resistance in the power formula equals _____.

(a) E x I   (b) E²/W   (c) E²I   (d) I²/W

7. Of the six ways of producing emf, which method is used the least?

(a) **pressure**   (b) **solar**   (c) **chemical action**   (d) **friction**

8. As the temperature increases, the resistance of most conductors also increases, except _____.

(a) **silver**   (b) **brass**   (c) **carbon**   (d) **zinc**

9. The hot resistance of a 100 watt incandescent bulb is about _____ times its cold resistance.

(a) **10**   (b) **2**   (c) **50**   (d) **100**

10. The purpose of load in an electrical circuit is to _____.

(a) **utilize electrical energy**   (b) **increase the current**
(c) **decrease the current**   (d) **none of these**

11. A device for making, breaking, or changing connections in a circuit under load is a _____.

(a) **inductor**   (b) **growler**   (c) **relay**   (d) **switch**

EXAM #2

12. What relationship determines the efficiency of electrical equipment?

(a) The power input divided by the output.
(b) The volt-amps x the wattage.
(c) The va divided by the pf.
(d) The power output divided by the input.

13. The conductance of a conductor is the ease in which current flows through it. It is measured in _____.

(a) teslas   (b) henrys   (c) mhos   (d) vars

14. In which of the following would a rheostat most likely not be used?

(a) transformer   (b) motor
(c) generator   (d) motor-generator set

15. Nichrome wire having a resistance of 200Ω per 1000 feet is to be used for a heater that requires a total resistance of 10Ω. The length of wire required would be_____ feet.

(a) 10   (b) 25   (c) 30   (d) 50

16. If the number of valence electrons of an atom is less than four, the substance is generally _____.

(a) a conductor   (b) a semiconductor
(c) an insulator   (d) none of these

EXAM #2

17. Electrical appliances are connected in parallel because it _____.

(a) makes the operation of appliances independent of each other
(b) results in reduced power loss
(c) is a simple circuit
(d) draws less current

18. E.M.F. in a circuit _____.

(a) maintains potential difference
(b) causes current to flow
(c) increases circuit resistance
(d) none of these

19. The resistance of a material is _____ its cross section area.

(a) independent of
(b) directly proportional to
(c) inversely proportional to
(d) none of these

20. The specific resistance of a conductor _____ with rise in temperature.

(a) remains the same
(b) increases
(c) decreases
(d) none of these

EXAM #3

1. What is the power factor of an incandescent bulb?

(a) unity   (b) 0.7 leading   (c) 0.7 lagging   (d) zero

2. Which of the following is NOT true about alternating current?

(a) develops eddy current losses
(b) it can be transformed
(c) is suitable for charging batteries
(d) interferes with communication lines

3. Metallic shielding is used on cables to _____.

(a) decrease thermal resistance
(b) decrease corona effect
(c) control the electrostatic voltage stress
(d) all of these

4. If three resistors are connected in series _____.

(a) the voltage across each is the same
(b) they must have a different resistance value
(c) the current across each is the same
(d) they must have the same resistance value

5. The specific resistance of a wire depends on _____.

(a) its material   **(b) its length**
(c) its csa   **(d) all of these**

6. Insulating materials have the function of _____.

(a) conducting very large currents
(b) preventing an open circuit
(c) storing very large currents
(d) preventing a short circuit between conducting wires

7. A no-load test is performed on a transformer for determining _____.

(a) efficiency of the transformer   **(b) copper loss**
(c) magnetizing current and loss   **(d) shorts**

8. Harmonics in a transformer leads to _____.

(a) increased core losses
(b) increased I²R losses
(c) interference in communication circuits
(d) all of these

9. In order to draw more current from a source _____.

(a) resistors are connected in series
(b) resistors are connected in parallel
(c) resistors are connected in series-parallel
(d) any of these

10. Power is defined as _____.

(a) the product of force and distance
(b) the rate of doing work
(c) the capacity of doing work
(d) the energy dissipated by a load

EXAM #3

11. The resistance of an open circuit is equal to ____.

(a) less than one Ω (b) zero
(c) infinity   (d) none of these

12. Inductance is measured in ____.

(a) henrys   (b) teslas   (c) webers   (d) mhos

13. Static electricity is generally produced by ____.

(a) chemical   (b) pressure   (c) heat   (d) friction

14. If the temperature of carbon is increased, its resistance will ____.

(a) double   (b) decrease   (c) increase   (d) remain same

15. The nucleus of an atom contains ____.

(a) protons and electrons
(b) protons and neutrons
(c) electrons and neutrons
(d) electrons, neutrons, and protons

16. If the resistance remains constant, and the current is doubled, the power is ____.

(a) halved   (b) multiplied by four
(c) doubled   (d) remains the same

EXAM #3

17. If a circuit is connected to allow current to flow in only one path, the circuit is connected in ____.

(a) delta   (b) salient   (c) parallel   (d) series

18. If two loads are connected in parallel with the source, connecting a third load in parallel in the circuit will ____.

(a) increase the total voltage
(b) increase the total current
(c) decrease the total voltage
(d) decrease the total current

19. The conductance (mho) of a circuit refers to the degree to which the circuit ____.

(a) opposes the rate of current flow
(b) permits or conducts voltage
(c) opposes the rate of voltage changes
(d) permits or conducts current flow

20. The minimum number of resistors in a series-parallel circuit would be ____.

(a) three   (b) two   (c) four   (d) five

EXAM #4

1. Dielectric strength is the _____.

(a) ability of the insulation to withstand the voltage
(b) strength of the magnetic field flux
(c) allowable current the conductor can carry
(d) ability of the insulator to tracking

2. In an alternating current circuit containing only resistance, the power factor would be closest to _____.

(a) 100%   (b) 80%   (c) 75%   (d) 50%

3. A load of one milliampere at one kilovolt would consume one _____.

(a) megawatt   (b) watt   (c) milliwatt   (d) kilowatt

4. A wire with a resistance of 0.491Ω per k/ft would have a resistance of _____ Ω for nine feet of wire.

(a) .004419   (b) 4.419   (c) .4419   (d) .04419

5. The current flow through a 10Ω resistor is 5 amps. The power consumed by this load would be _____ watts.

(a) 50   (b) 2   (c) 500   (d) 250

6. The voltage induced in a coil of wire is greatest where the magnetic field within the coil is _____.

(a) increasing   **(b) decreasing**
(c) constant   **(d) changing most rapidly**

7. A voltmeter is connected across the positive and negative terminals of a battery. The difference between a no-load voltage and a voltage when current is drawn from the battery is the _____.

(a) voltage of the battery
(b) voltage drop across the load resistance
(c) internal voltage drop in the battery
(d) external voltage drop of the battery

8. A 100 watt light bulb operating for 40 hours will consume _____ kwh.

(a) 4   (b) 40   (c) 4000   (d) 400

9. A DC circuit generally has _____ as the load.

(a) inductance   **(b) resistance**
(c) capacitance   **(d) impedance**

10. Inductance and capacitance are not relevant in a DC circuit because _____.

(a) they never exist in DC   **(b) DC is a simple circuit**
(c) DC frequency is zero   **(d) none of these**

EXAM #4

11. The hot resistance of a 100 watt bulb @ 120v is _____ Ω.

(a) 144   (b) .83   (c) 1.2   (d) 25

12. When a number of resistances are connected in parallel, the total resistance will always be _____.

(a) greater than the smallest resistance
(b) greater than the largest resistance
(c) less than the smallest resistance
(d) none of these

13. A load 6Ω and a 3Ω are connected in parallel. The total resistance would be ___ Ω.

(a) 9   (b) 2   (c) 18   (d) 1

14. Two 100w @ 120 volt light bulbs are connected parallel in the circuit. The total resistance would be _____ Ω.

(a) 288   (b) 144   (c) 72   (d) 36

15. Cells are connected in series when _____ is required.

(a) higher current
(b) higher voltage
(c) both high current and voltage
(d) none of these

EXAM #4

16. A positively charged body has _____.

(a) deficit of electrons   (b) excess of electrons
(c) deficit of neutrons   (d) excess of neutrons

17. Paper does not exhibit electricity because it contains the same number of _____.

(a) neutrons and protons   (b) protons and electrons
(c) neutrons and electrons   (d) none of these

18. Electrons in the last orbit of an atom are called _____ electrons.

(a) free   (b) bound   (c) valence   (d) loose

19. Which of the following would have the highest dielectric strength?

(a) oiled paper   (b) air   (c) mica   (d) glass

20. The unit of flux density is called the _____.

(a) maxwell   (b) tesla   (c) weber   (d) gauss

EXAM #5

1. The voltage drop across the 15Ω resistor is _____ volts.

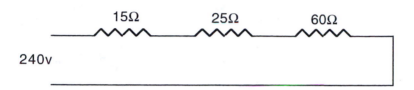

2. The total circuit resistance would be _____ ohms.

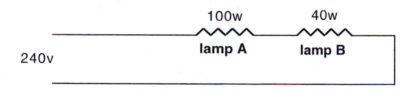

3. In the circuit shown above _____.

(a) lamp A will be brighter than lamp B
(b) lamp B will be brighter than lamp A
(c) lamp A and B would have the same brightness
(d) none of these

EXAM #5

4. The current flow in the 2.5Ω resistor is _____ amps.

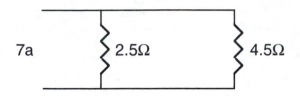

5. The resistance across terminals A and B would be _____ Ω.

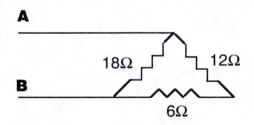

6. In the circuit above, if the applied voltage is 180, the current flowing in the:

6Ω resistor would be _____ amps.

12Ω resistor would be _____ amps.

18Ω resistor would be _____ amps.

Total circuit current would be _____ amps.

EXAM #5

7. The resistance across terminals A and C would be _____ Ω.

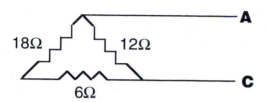

8. What is the current flow in the parallel circuit?

_____ amps.

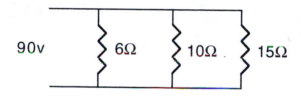

9. A motor with 50 amps, 230 volts at 80% power factor would deliver approximately _____ horsepower.

10. The resistance across terminals A and B is _____ ohms.

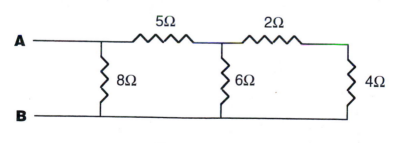

EXAM #6

1. The applied voltage to this parallel circuit is _____ volts.

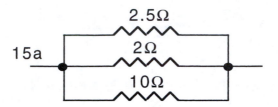

2. What is the voltage drop at the 7Ω resistor in the circuit shown below?

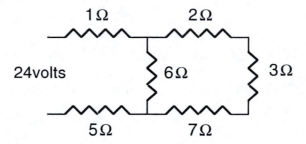

3. What is the current flow in the 4Ω resistor?

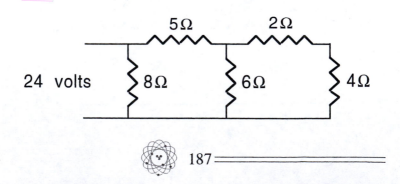

EXAM #6

4. The total resistance of this circuit is _____ ohms.

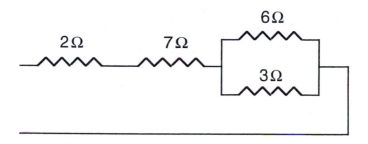

5. What is the total current flow in the circuit shown below?

_____ amps.

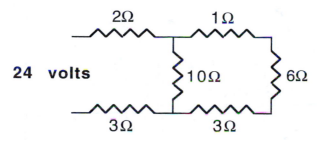

6. The resistance across terminals L2 and L3 is _____ Ω.

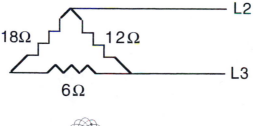

EXAM #6

7. The applied voltage in the circuit below would be _____.

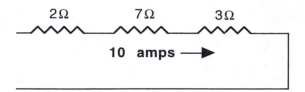

8. The resistance across terminals L2 and L1 is _____ Ω.

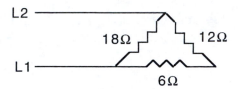

9. The applied voltage in the circuit below is _____ volts.

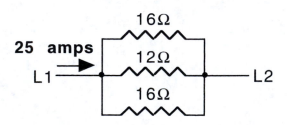

10. The current flow in the 9Ω resistor would be _____ amps.

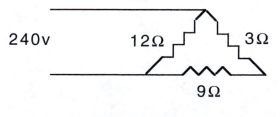

EXAM #7

1. The resistance at terminals L1 and L3 would be _____Ω.

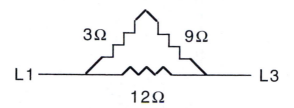

2. If a voltage of 120 is applied across terminals L1 - L3 in the circuit shown above, the current through the 12Ω resistor would be _____ amps.

3. The resistance at terminals L1-L2 would be _____ Ω.

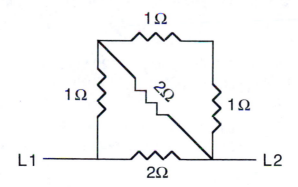

EXAM #7

4. The resistance at terminals L1 - L2 is ____ ohms.

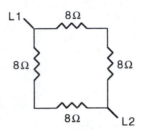

5. A 20Ω load with 5 amps current flow would use ____ watts of power.

6. How much power is consumed in a circuit which operates at 240 volts, draws 8 amperes, and has a power factor of 80%?

7. Measurements were taken in an AC circuit and the current flowing was 10 amps; the voltage 240. The wattmeter reads 1800 watts. The power factor of this circuit would be ____%.

8. A 100 watt light bulb is rated @ 120 volts. What is the wattage if the applied voltage is 115 volts?

9. Two 100 watt lamps connected in series across a 120 volt line draws .42 amperes. The total power consumed is ____ watts.

10. A 15Ω resistance draws 15 amps of current. The watts consumed would be ____.

EXAM #8

1. What is the total wattage of the circuit shown below?

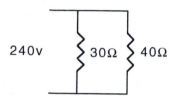

2. What is the total wattage of the 240 volt circuit below?

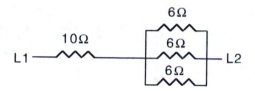

3. What is the wattage of the 10Ω resistor in the circuit above?

4. What is the wattage of one 6Ω load?

5. What is the total wattage of the three 6Ω loads?

6. What is the total wattage of the five loads in the 120 volt circuit?

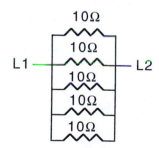

EXAM #8

7. What is the total power of these four loads?

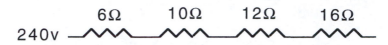

6Ω   10Ω   12Ω   16Ω

240v

8. What is the total power of these four light bulbs?

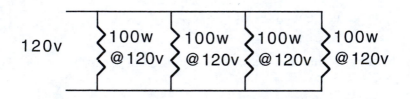

100w   100w   100w   100w
@120v   @120v   @120v   @120v

120v

9. What is the total power of these four light bulbs?

120v

100w   100w   100w   100w
@120v   @120v   @120v   @120v

10. What is the total wattage of these two light bulbs?

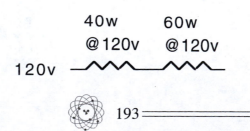

40w   60w
@120v   @120v

120v

EXAM #9

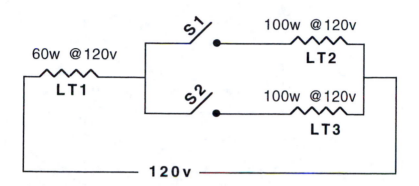

1. With switches S1 and S2 closed the total circuit resistance would be _____ ohms.

2. With switch S1 open and S2 closed the circuit resistance would be _____ ohms.

3. With switches S1 and S2 closed the total circuit current would be _____ amps.

4. With switch S1 open and S2 closed the circuit current would be _____ amps.

5. With switches S1 and S2 closed the voltage drop at the 60 watt light would be _____ volts.

EXAM #9

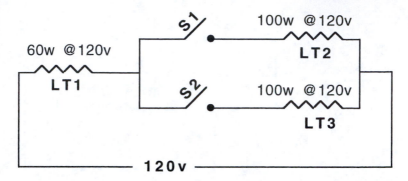

6. With switches S1 open and S2 closed the voltage drop at the 60 watt light would be _____ volts.

7. With switches S1 and S2 closed the voltage drop at LT2 would be _____ volts.

8. With switches S1 open and S2 closed the voltage drop at LT3 would be _____ volts.

9. With switches S1 open and S2 closed the wattage at the 60 watt light would be _____ watts.

10. With switches S1 and S2 closed the total circuit wattage would be _____ watts.

EXAM #10

1. A 40w and a 75w light bulb are connected in series. Which bulb will give off the most wattage?

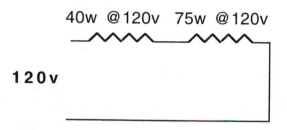

2. In the circuit shown below what is the current flow in the 16Ω load?

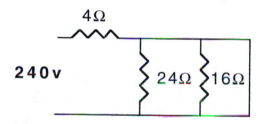

3. In the circuit below what is the total circuit resistance?

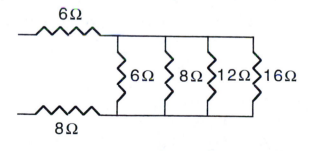

EXAM #10

4. A ceiling fan in a residence draws .65 amps @ 120v has four 40 watt light bulbs rated @ 120 volts. The total circuit wattage would be _____ watts.

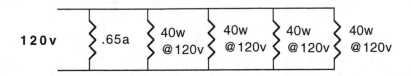

5. The total current flow in the circuit above with everything turned on would be _____ amps.

6. What is the total circuit resistance in the circuit above?

7. What would be the voltage drop at the 120 volt leads supplying the fan?

EXAM #10

8. What is the wattage of the 6Ω resistor in the circuit shown below?

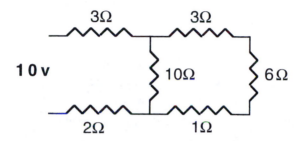

9. What is the resistance of R3 if all the resistors are of equal value and the current flow through the circuit is 5 amps?

10. The total circuit resistance is 12Ω. What is the resistance of R2?

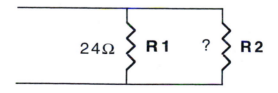

FINAL EXAM

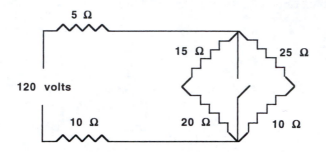

1. What is the total resistance of this 120 volt circuit with the switch open as shown?

(a) **85 ohms** (b) **17.5 ohms** (c) **32.5 ohms** (d) **95 ohms**

2. In the circuit shown below, the ammeter would read _____ amps.

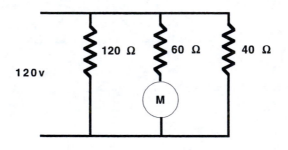

(a) **2** (b) **6** (c) **60** (d) **.545**

FINAL EXAM

3. In the circuit shown below all three resistors have the same ohmic value, the resistance of R2 would be _____ ohms.

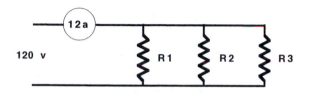

(a) 10   (b) 20   (c) 30   (d) 40

4. In the circuit shown below, the source voltage would be _____ volts.

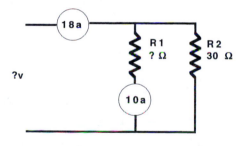

(a) 58   (b) 120   (c) 90   (d) 240

5. A nichrome wire having a resistance of 250 ohms per 100 feet is used to wire a toaster. What would be the total length of wire if the toasters total resistance is 10 ohms?

(a) 4 feet   (b) 5 feet   (c) 25 feet   (d) none of these

 200

FINAL EXAM

6. Resistance A-B is lowest when switches _____ are open.

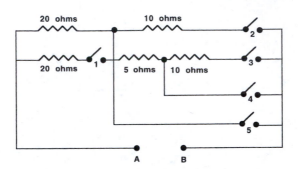

(a) 1 through 5   **(b) 2 through 5**
(c) 2, 3, and 4   **(d) 3, 4, and 5**

7. What is the current flowing in this circuit with the switch closed?

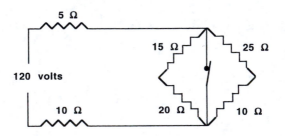

(a) 1.41 amps   (b) 1.71 amps   (c) 3.69 amps   (d) 8 amps

FINAL EXAM

8. In the circuit shown below, if all three resistors are rated at 250 watts, which resistor or resistors would overheat?

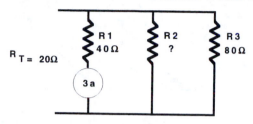

(a) R1   (b) R2   (c) R3   (d) none of them

9. What is the total resistance in the series-parallel circuit?

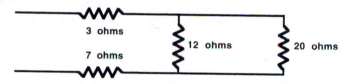

(a) 42Ω   (b) 4.28Ω   (c) 17.5Ω   (d) none of these

10. With only switch 4 closed and a line voltage of 225 volts, the drop across one of the 10 ohm resistors is _____ volts.

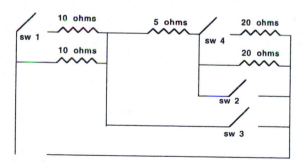

(a) 225   (b) 90   (c) 64.3   (d) 56.3

FINAL EXAM

SUMMARY

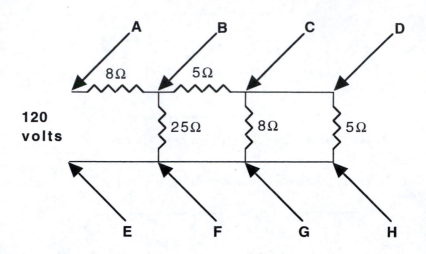

Solve the following for the circuit above:

AB = _____ volts   AB = _____ amps

AE = _____ volts   BC = _____ amps

BC = _____ volts   BF = _____ amps

CD = _____ volts   DH = _____ amps

DH = _____ volts   CG = _____ amps

BF = _____ volts   Total circuit _____ amps

FG = _____ volts

ANSWERS

ANSWERS

EXAM #1

1. (b)	11. (c)
2. (a)	12. (c)
3. (c)	13. (b)
4. (a)	14. (d)
5. (c)	15. (c)
6. (d)	16. (a)
7. (a)	17. (b)
8. (d)	18. (a)
9. (b)	19. (a)
10. (c)	20. (d)

EXAM #2

1. (c)	11. (d)
2. (c)	12. (d)
3. (c)	13. (c)
4. (a)	14. (a)
5. (b)	15. (d)
6. (b)	16. (a)
7. (d)	17. (a)
8. (c)	18. (a)
9. (a)	19. (c)
10. (a)	20. (b)

EXAM #3

1.	(a)	11.	(c)
2.	(c)	12.	(a)
3.	(d)	13.	(d)
4.	(c)	14.	(b)
5.	(d)	15.	(b)
6.	(d)	16.	(b)
7.	(c)	17.	(d)
8.	(d)	18.	(b)
9.	(b)	19.	(d)
10.	(b)	20.	(a)

EXAM #4

1.	(a)	11.	(a)
2.	(a)	12.	(c)
3.	(b)	13.	(b)
4.	(a)	14.	(c)
5.	(d)	15.	(b)
6.	(d)	16.	(a)
7.	(c)	17.	(b)
8.	(a)	18.	(c)
9.	(b)	19.	(d)
10.	(c)	20.	(b)

EXAM #5 Answers

1. VD = I x R   The total resistance is $15\Omega + 25\Omega + 60\Omega = 100\Omega$.   Next find the current flow in the circuit. I = E/R = 240v/100Ω = 2.4 amps flowing through the series circuit. 2.4a x 15Ω = **36 volts** is dropping at the 15Ω.

2. R = E²/W   240v x 240v/100w = 576Ω
                    240v x 240v/40w = 1440Ω
Resistance adds in series 576Ω + 1440Ω = **2016 R total.**

3. Find the current flow in this series circuit   I = E/R. 240v/2016Ω = .119 amps W = E x I.   Find the voltage at each load. E = I x R   .119a x 576Ω = 69v at the 100w bulb. The voltage at the 40w bulb would be .119a x 1440Ω = 171v.
Checkpoint 69v + 171 = 240 the applied voltage. The wattage at each bulb in this *series* connected circuit would be W = E x I. The 100w bulb connected in series would have a wattage of 69v x .119a = 8.2 watts. The 40 watt bulb in series would have a wattage of 171v x .119a = 20.3 watts. **Lamp B would be brighter with 20 watts.**

4. To solve this question we need another *known*. Find the voltage of this circuit. E = I x R. The amps is , solve R. Total resistance in parallel is less than the smallest resistance.

$\overline{R1 \times R2}$ = $\overline{2.5\Omega \times 4.5\Omega} = \overline{11.25}$ = 1.6 Rtotal
$R1 + R2$ 2.5Ω + 4.5Ω   7
E = 7a x 1.6Ω = 11.25 applied voltage.
I = E/R = 11.25v/2.5Ω = **4.5 amps** flowing in the 2.5Ω.

EXAM #5 Answers

5. This circuit is connected series-parallel. As the applied voltage is at the 18Ω and the 12Ω and 6Ω are connected in series. The circuit actually looks like this:

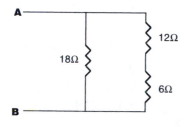

The 12Ω and 6Ω are connected in series so the resistance would add for a total of 18Ω. Now the circuit is two 18Ω resistors in parallel.

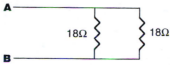

The resistance at terminals A and B would be 18Ω/2 = **9Ω.**

6. The 6Ω is in series with the 12Ω for a resistance of 18Ω. I = E/R = 180v/18Ω = **10 amps** flowing in the 6Ω resistor.
The current flow in the 12Ω in the same in series, **10amps.**
The current flow in the 18Ω is 180v/18Ω = **10 amps.**

The total current would be 180v/9Ω = **20 amps.**

The 20 amps divides in the circuit as shown below:

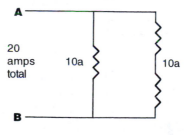

EXAM #5 Answers

7. This circuit is connected series-parallel. As the applied voltage is at the 12Ω and the 18Ω and 6Ω are connected in series. The circuit actually looks like this:

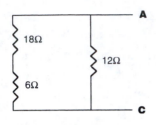

The 18Ω and 6Ω are connected in series so the resistance would add for a total of 24Ω. Now the circuit is two resistors in parallel a 24Ω and a 12Ω.

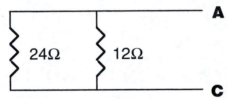

The total resistance for unequal resistors is:

$$\frac{R1 \times R2}{R1 + R2} \qquad \frac{24Ω \times 12Ω = 288}{24Ω + 12Ω = \ \ 36} \qquad = \qquad \textbf{8 ohms.}$$

8. Current adds together in parallel. $I = E/R$

90v/6Ω = 15 amps   90v/10Ω = 9a   90v/15Ω = 6a

15a + 9a + 6a = **30 amps** total

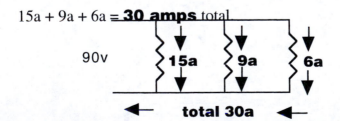

EXAM #5 Answers

9. 50a x 230v = 11,500 x 80% PF = 9200/746w = **12 hp.**

10. Start at the end of the circuit with the 2Ω nd 4Ω which add in series making a 6Ω. This 6Ω is now in parallel with the other 6Ω. The circuit now looks like this:

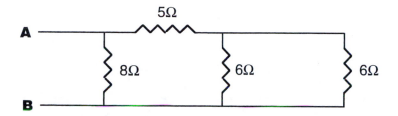

The two 6Ω are in parallel making a combined resistance of 6Ω/2 = 3Ω. Now the circuit is reduced to this form:

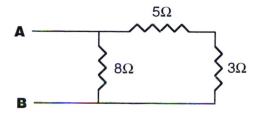

The 5Ω and 3Ω are in series and add to make an 8Ω resistor.

Now the circuit has two 8Ω resistors in parallel and looks like this:

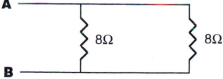

The resistance at terminals A and B is 8Ω/2 = **4 ohms.**

EXAM #6 Answers

1. $E = I/R$ Find the total resistance of the unequal parallel loads. $\dfrac{\text{R1 x R2}}{\text{R1 + R2}} = \dfrac{10\Omega \times 2.5\Omega}{10\Omega + 2.5\Omega} = \dfrac{25}{12.5} = 2$

$\dfrac{\text{R3 x 2}\Omega}{\text{R3 + 2}\Omega} = \dfrac{2\Omega \times 2\Omega}{2\Omega + 2\Omega} = \dfrac{4\Omega}{4\Omega} = 1\Omega$ total resistance

$E = I \times R$ $= 15a \times 1\Omega = $ **15 volts** applied.

2. $VD = I \times R$ First step is to solve R, the total resistance. The 2Ω, 3Ω, and 7Ω are connected in series with the resistance combining for a total of 12Ω. The 12Ω is now in parallel with the 6Ω with the circuit now looking like this:

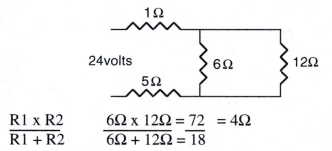

$\dfrac{\text{R1 x R2}}{\text{R1 + R2}}$ $\dfrac{6\Omega \times 12\Omega}{6\Omega + 12\Omega} = \dfrac{72}{18} = 4\Omega$

Now the combined circuit looks like this:

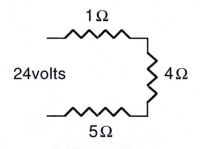

EXAM #6 Answers

The three resistors 1Ω, 4Ω, and 5Ω add together in series for a total circuit resistance of 10Ω.

I = E/R   **24v/10Ω = 2.4 amps current flow.**

The voltage drop is I x R.

2.4a x 1Ω = 2.4VD
                    1Ω

2.4a x 4Ω = 9.6VD
4Ω

5Ω
2.4a x 5Ω = 12VD

2.4v + 9.6v + 12v = 24 volts applied.

Solve the current flowing through each resistor:

I = E/R   9.6v/6Ω = 1.6a flow through 6Ω resistor.
2.4a - 1.6a = .8a flow through 2Ω, 3Ω, and 7Ω. 1Ω and 5Ω have the 2.4a current flow.

The 2.4 amps of current flow will divide as shown below:

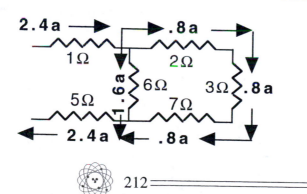

EXAM #6 Answers

Now you can solve the voltage drop at any resistor in the circuit:

$$VD = I \times R$$

.8a x 2Ω = 1.6vd
.8a x 3Ω = 2.4vd
.8a x **7Ω = 5.6vd**

9.6 total voltage drop

3. First step is to solve R, the total resistance. The 2Ω and 4Ω are connected in series with the resistance combining for a total of 6Ω. The 6Ω is now in parallel with the 6Ω with the circuit now looking like this:

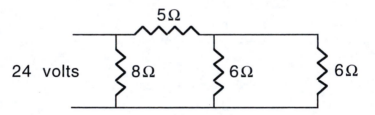

Since the two 6Ω resistors in parallel are equal the resistance would be 6Ω/2 = 3Ω. Now the circuit is reduced to this form:

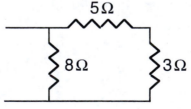

The 5Ω and 3Ω add in series for a resistance of 8Ω which is in parallel with the 8Ω resistor.

EXAM #6 Answers

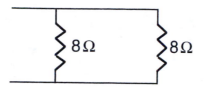

The two are equal parallel, 8Ω/2 = 4Ω **total resistance.**

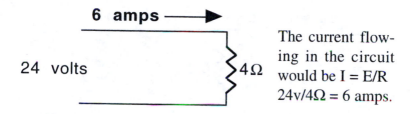

The current flowing in the circuit would be I = E/R
24v/4Ω = 6 amps.

Solve the current flowing through each resistor:

I = E/R   24v/8Ω = 3a flow through 8Ω resistor.
6a - 3a = 3a flows through the 5Ω.

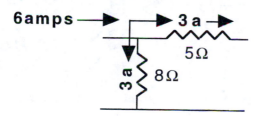

Now the 3 amps are flowing towards the end of the circuit and must divide into two paths.

EXAM #6 Answers

The 2Ω and 4Ω combine to a 6Ω in parallel with the other 6Ω.

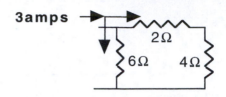

The 3 amps divides equal 1.5 amps.

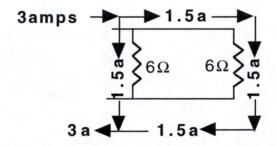

The 6 amps of current flow will divide and join as shown below:

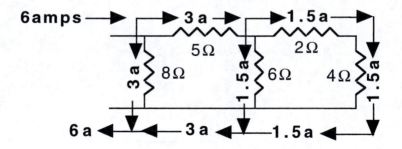

The answer is the current flow in the 4Ω resistor is **1.5a.**

EXAM #6 Answers

4. Start at the end of the circuit with the 6Ω and 3Ω that are connected in parallel and combine them into one resistor.

$$\frac{R1 \times R2}{R1 + R2} \quad \begin{array}{l} = 6\Omega \times 3\Omega = \dfrac{18}{9} = 2\Omega \\ = 6\Omega + 3\Omega = \end{array}$$

Now the circuit looks like this:

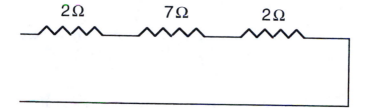

In series the resistances add together: $2\Omega + 7\Omega + 2\Omega =$ **11Ω total circuit resistance.**

5. $I = E/R$ The first step is to solve the total circuit resistance. The 1Ω, 6Ω, and 3Ω are connected in series and add together for a total of 10Ω. This 10Ω would be in parallel with the other 10Ω. These would combine $10\Omega/2 = 5\Omega$. Now the three loads are in series and add together for a total resistance of 10Ω.

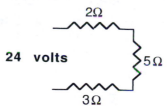

I total = E total/R total = $24v/10\Omega =$ **2.4 amps total.**

EXAM #6 Answers

6. The 18Ω and 6Ω are connected in series and would add together for a total of 24Ω. The 24Ω would now be in parallel with the 12Ω. Now the circuit would look like this:

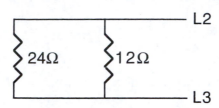

$$\frac{R1 \times R2}{R1 + R2} \quad = \frac{24Ω \times 12Ω}{24Ω + 12Ω} = \frac{288}{36} = \textbf{8Ω resistance.}$$

7. E = I x R   The loads are connected in series so the resistance would add together 2Ω + 7Ω + 3Ω = 12Ω total. 10 amps x 12 ohms = **120 volts applied.**

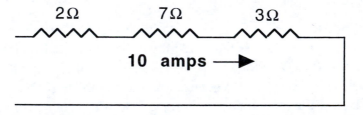

8. The 12Ω and 6Ω are connected in series adding for a total of 18Ω. The 18Ω is now in parallel with the 18Ω. 18Ω/ 2 =
9Ω resistance.

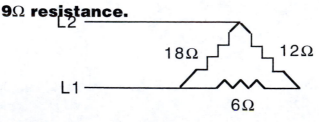

EXAM #6 Answers

9. $E = I \times R$ First find the total resistance.
The two 16Ω would combine for and equal resistance of
$16\Omega/2 = 8\Omega$. This 8Ω would be in parallel with the 12Ω.

$$\frac{R1 \times R2}{R1 + R2} \quad \begin{array}{l} = 8\Omega \times 12\Omega = 96 = 4.8\Omega \text{ total resistance.} \\ = \overline{8\Omega + 12\Omega} = \overline{20} \end{array}$$

25 amps x 4.8 ohms = **120 volts applied.**

10. $I = E/R$ Find the total circuit resistance first. The 3Ω
and 9Ω are connected in series adding together for a total
of 12Ω. The 12Ω is in parallel with the 12Ω combining for
a total resistance of $12\Omega/2 = 6$ ohms.

The total current flow in the circuit would be 240v/6Ω =
40 amps.

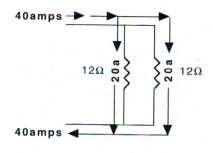

The 40a current would
divide equally between
the two loads.

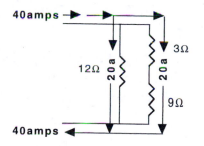

The current flow through
the 9Ω resistor would be
20 amps.

EXAM #7 Answers

1. The 3Ω and 9Ω are connected in series adding together for a total resistance of 12Ω. The 12Ω would be in parallel with the 12Ω. The two would combine 12Ω/2 = **6Ω at L1-L3.**

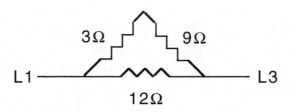

2. I = E/R   120v/ 6Ω = 20 amperes.

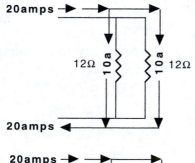

The 20a current would divide equally between the two loads.

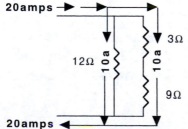

The current flow through the 12Ω resistor would be **10 amps.**

EXAM #7 Answers

3. The two 1Ω are connected in series adding together for a combined resistance of 2Ω. The 2Ω is in parallel with the 2Ω for a combined resistance of 2Ω/2 = 1Ω.

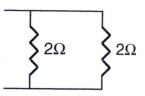

The two 1Ω in series add together for a total of 2Ω which is in parallel now with the 2Ω combining for a total resistance of 2Ω/2 = **1Ω.**

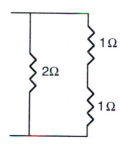

4. The two 8Ω are connected in series adding to a combined resistance of 16Ω. The circuit looks like this:

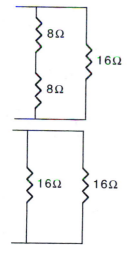

The two 8Ω are connected in series adding to a combined resistance of 16Ω. The 16Ω is now in parallel with the 16Ω for a combined resistance of 16Ω/2 = **8Ω at L1 - L2.**

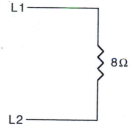

EXAM #7 Answers

5. $W = I^2R$ 5a x 5a x 20Ω = **500 watts.**

6. $W = E \times I \times PF$ 240v x 8a x 80% = **1536 watts.**

7. $PF = W/va$ 1800w/240v x 10a = **75% PF.**

8. The bulb is rated 100w @ 120v. The first step is to find the fixed resistance. $R = E^2/W$ = 120v x 120v/100w = 144Ω.
To find the wattage at another voltage: $W = E^2/R$
115vx115v/144Ω = **91.84 watts.**

9. $W = E \times I$ 120v x .42 amps = **50 watts.**

10. $W = I^2R$ 15a x 15a x 15Ω = **3375 watts.**

EXAM #8 Answers

1. Sometimes you can solve the question in more than one way.

$W = E \times I$ First find I of the parallel circuit. $I = E/R$.
240v/30Ω = 8a   240v/40Ω = 6a   Current adds
together in parallel. 8a + 6a = 14a total.
240v x 14a = 3360 watts.

Or you can solve it by $W = E^2/R$. To find the resistance in parallel:

$$\frac{R1 \times R2 = 30Ω \times 40Ω = 1200}{R1 + R2 = 30Ω + 40Ω = \quad 70} = 17.14Ω$$

240v x 240/17.14Ω = 3360 watts.

2. Find the circuit resistance first. The three 6Ω in parallel would combine (6Ω/3) for a resistance of 2Ω which would add in series to the 10Ω for a total resistance of 12Ω.
$W = E^2/R$ 240v x 240v/12Ω = **4800 watts.**
$I = E/R$ 240v/12Ω = 20 amps of current flow.
$W = E \times I$ 240v x 20a = 4800 watts.

3. $W = I^2R$ 20a x 20a x 10Ω = **4000 watts.**

EXAM #8 Answers

4. The 20a current flow will divide equally among the three 6Ω resistors connected in parallel. 20a/3 = 6.6666666 amps.

$W = I^2R$ 6.6666666a x 6.6666666a x 6Ω = 266.66665 or **267 watts.**

Checkpoint: 266.66665Ω x 3 resistors = 800 watts.

Another way to solve this calculation is W = E x I. First we would solve E = I x R:

The combined circuit resistance is 12Ω. The current flow is 20 amps (E/R). The vd at the 10Ω is 200 volts (20a x 10Ω).
The vd at the 2Ω is 40 volts (20a x 2Ω).

The 2Ω is a combination resistance of the three 6Ω in parallel. The voltage applied to these three 6Ω resistors is 40 volts.

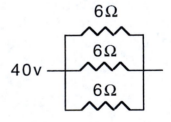

W = E x I = 40v x 6.6666666a = 266.66666 or 267w.

EXAM #8 Answers

5. $W = I^2R$ 20a x 20a x 2Ω = **800 watts.**

6. $W = E^2/R$ The total resistance in parallel would be (10Ω/5) 2 ohms. 120v x 120v/2Ω = **7200 watts.**

7. $W = E^2/R$ The total resistance in series would be 6Ω + 10Ω + 12Ω + 16Ω = 44Ω.   240v x 240v/44Ω = **1309 watts.**

8. $W = E^2/R$ First find the fixed resistance. $R=E^2/W$ 120v x 120v/100w = 144Ω. Resistance adds in series for a total of 576 ohms. 120v x 120v/576Ω = **25 watts.**

9. $W = E^2/R$ First find the fixed resistance. $R=E^2/W$ 120v x 120v/100w = 144Ω. The total resistance in parallel would be (144Ω/4) = 36Ω. 120v x120v/36Ω =**400 watts.** Wattage adds in parallel 100w+100w+100w+100w = 400w.

10. $W = E^2/R$ First find the fixed resistance. $R=E^2/W$ 120v x 120v/40w = 360Ω. 120v x 120v/60w = 240Ω.Resistance adds in series for a total of 600 ohms. 120v x 120v/600Ω = **24 watts.**

EXAM #9 Answers

1. First step find the total circuit resistance. $R = E^2/W$ 120v x 120v/100w = 144Ω. The two 100w lights are in parallel for a combined resistance of 72Ω (144Ω/2). The other light would have a resistance of 120v x 120v/60w = 240Ω. The 240Ω and the 72Ω would add together for a total circuit resistance of **312 ohms.**

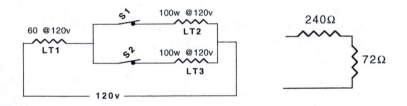

2. With S1 open the 60 watt and 100 watt are connected in series. The resistance would add 240Ω + 144Ω = **384Ω.**

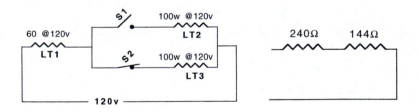

3. $I = E/R$ = 120v/312Ω = **.3846153 amps.**

EXAM #9 Answers

4. $I = E/R$ $= 120v/384\Omega =$ **.3125 amps.**

5. $E = I \times R = .3846153a \times 240\Omega =$ **92.3 volts dropped.**

6. $E = I \times R = .3125a \times 240\Omega =$ **75 volts dropped.**

7. $E = I \times R$ The current in parallel would divide equally between the two 100 watt bulbs .3846153a/2 = .1923076a. .1923076a x 144$\Omega =$ **27.7 volts dropped at LT2.**

Checkpoint: 92.3 vd + 27.7 vd = 120 volts applied.

8. $E = I \times R$ $= .3125a \times 144\Omega =$ **45 volts dropped.**

Checkpoint: 75 vd + 45 vd = 120 volts applied.

9. $W = E^2/R$ $= 92.3v \times 92.3v/240\Omega =$ **35.49 watts.**

10. $W = E \times I$ $= 120v \times .3846153a =$ **46.15 watts.**

Checkpoint:
$W = I^2R$
.3846153a x .3846153a x 312Ω = 46.15 watts.

$W = E^2/R$
120v x 120v/312Ω = 46.15 watts.

EXAM #10 Answers

1. First find the circuit resistance $R = E^2/W$.
120v x 120v/40w = 360Ω.   120v x 120v/75w = 192Ω.

Resistance adds in series 360Ω + 192Ω = 552Ω total.
Next find the current flow in the series circuit $I = E/R$.
120v/552Ω = .2173913 amps flowing through circuit.
$W = I^2R$ = .2173913a x .2173913a x 360Ω = 17 watts.
.2173913a x .2173913a x 192Ω = 9 watts.

The **40w bulb** would give off more wattage at 17 watts than the 75w bulb at 9 watts.

In series the current flow in the entire circuit is the same .2173913 amps. The 40w bulb has a higher resistance at 360 ohms so it would take more pressure (voltage) to push the current through the 40w bulb.

Check the voltage drop in the circuit: $VD = I \times R$.

$$\longrightarrow \text{.2173913 amps} \longrightarrow$$
$$360\Omega \qquad 192\Omega$$
$$120v \;—\!\!\text{/\\/\\/\\}\!\!—\!\!\text{/\\/\\/\\}\!\!—\; 0v$$
$$\textbf{78.26vd} \qquad \textbf{41.74vd}$$

The circuit uses up 78.26 volts to push the current through the 360Ω bulb leaving only 41.74 volts left to push the same current through the 192Ω bulb. The wattage at the 75 watt bulb would be $W = E^2/R$ = 41.74v x 41.74v/192Ω = 9 watts.

Checkpoint: The total voltage drop of a circuit will equal the applied voltage. 78.26vd + 41.74 vd = 120v applied.

EXAM #10 Answers

2. I = E/R   First find the total circuit resistance. The 24Ω and 16Ω are connected in parallel.

$$\frac{R1 \times R2}{R1 + R2} \qquad \frac{24\Omega \times 16\Omega}{24\Omega + 16\Omega} = \frac{384}{40} \quad = 9.6\Omega$$

The 9.6Ω would add in series with the 4Ω for a total of 13.6 ohms total circuit resistance.

I = 240v/13.6Ω = 17.65 amps.

The circuit looks like this:

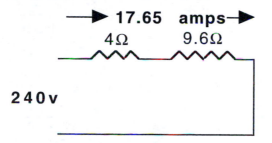

The current would divide into the 3 resistors. I = E/R. Next find the voltage drop at each resistor. VD = I x R.
17.65a x 9.6Ω = 169.44vd   17.65 x 4Ω = 70.6vd

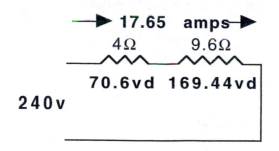

EXAM #10 Answers

$I = E/R$ 169.44v/24Ω = 7.06a
          169.44v/**16Ω = 10.59a**
                  17.65 total current flow.

          70.6v/4Ω   = 17.65a

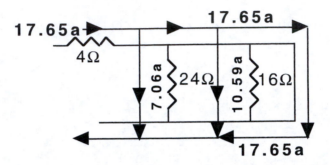

3. The 6Ω, 8Ω, 12Ω, and 16 Ω are connected in parallel.

6Ω x 8Ω = 48 = 3.43Ω   3.43Ω x 12Ω = 41.16 = 2.67Ω
6Ω + 8Ω = 14   3.43Ω + 12Ω = 15.43

2.67Ω x 16Ω = 42.72 = 2.28Ω
2.67Ω + 16Ω = 18.67

The 2.28Ω would add in series with the 6Ω and 8Ω for a total circuit resistance of **16.28 ohms.**

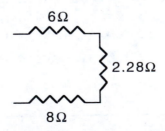

EXAM #10 Answers

4. The four 40 watt bulbs would add together in parallel for 160 watts. The fan motor wattage would be 120v x .65a = 78 watts for a total of 160w + 78w = **238 watts.**

5. I = W/E   40w/120v = .33a x 4 bulbs = 1.32a + .65a
fan
= **1.97 amps.**

6. R = E²/W   40 watt bulb = 120v x 120v/40w = 360Ω
   R = E/I   fan = 120v/.65a = 184.6 or 185Ω

Now the circuit looks like this:

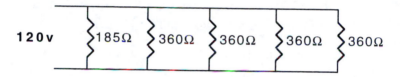

120v   185Ω   360Ω   360Ω   360Ω   360Ω

The 360Ω resistors are equal and would combine for a resistance of 360Ω/4 = 90Ω. The 90Ω would be in parallel with the fan at 185Ω.

$$\frac{185Ω \times 90Ω}{185Ω + 90Ω} = \frac{16650}{275} = \textbf{60.5 ohms total}$$

7. VD = I x R .65a x 184.6Ω = **120 volts.**

•Always remember all the voltage is used to push the .65 amps through the 60.5 ohm resistance. It is used in the form of wattage. The voltage is replaced every 1/60th of a second and used up again, again, and again. It drops so fast and is replaced so rapidly your voltmeter never indicates any sign of drop, it keeps on reading 120 volts all day long.

EXAM #10 Answers

8. First find the total resistance of the circuit. The 3Ω, 6Ω, and 1Ω add together for a resistance of 10Ω which is in parallel with the 10Ω combining for a total of 10Ω/2= 5Ω.

Now the circuit looks like this:

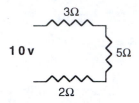

The 3Ω, 5Ω, and 2Ω add together for a total circuit resistance of 10Ω. The current flow would be I = E/R = 1 amp.

The voltage drop at each resistor would be VD = I x R =
1a x 3Ω = 3vd   1a x 5Ω = 5vd   1a x 2Ω = 2vd
The 1 amp current will divide equally as shown below.

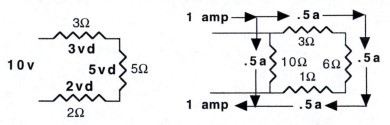

The voltage drop at each resistor would be VD = I x R =
.5a x 3Ω = 1.5vd   .5a x 6Ω = 3vd   .5a x 1Ω = .5vd

To find the wattage of the 6Ω resistor: W = E x I =
3v x .5a = **1.5 watts.**

EXAM #10 Answers

9. R = E/I   100v/5a = 20Ω total circuit resistance. The four resistors are connected in parallel and have equal value. The total resistance of the four is 20Ω. To find the resistance of R3 take 20Ω x 4 resistors = **80Ω.**

10. 24Ω/2 = 12Ω total.   R2 = **24 ohms.**

FINAL EXAM Answers

1. The circuit with the switch open looks like this:

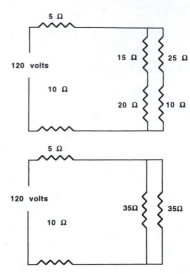

Start at the end of the circuit by combining:

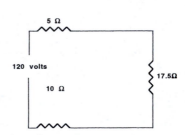

Now you have two 35Ω in
parallel which would be
35Ω/2 = 17.5Ω.

Now you add the loads together for the total resistance: 5Ω
+ 17.5Ω + 10Ω = **32.5 Ω total.**

FINAL EXAM Answers

2. This is a parallel circuit with unequal resistance. The current flow will be different in each load. To find the current the current meter is reading use I = E/R.

120v/60Ω = **2 amps**

3. For resistances of equal value in parallel, divide the resistance of one by the number of resistors. The total resistance is 10Ω, the resistance of each resistor would be 30Ω since they are equal.

R = E/I   120v/12a = 10 ohms

$$\frac{\text{Resistance of one}}{\text{Number of resistances}} = \frac{30}{3} = 10Ω$$

The resistance of R2 would be **30 ohms.**

4. E = I x R   = 8a x 30Ω = **240 volts**

5. $\dfrac{250 \text{ ohms}}{100 \text{ feet}}$ =   2.5 ohms per foot

$\dfrac{10 \text{ ohms total resistance}}{2.5 \text{ ohms per foot}}$ =   **4 feet of wire**

 234

FINAL EXAM Answers

6. When switches 2, 3, and 4 are open, switches 1 and 5 are closed. With switch 5 closed the circuit has a resistance of 20 ohms. Switch 1 is also closed but does not add any resistance to the circuit.

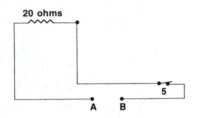

The lowest resistance is when switches **2, 3, and 4 are open.**

7. $I = E/R$ $120v/15\Omega$ =   **8 amps**

8. $E = I \times R$ 3 amps x 40 ohms = 120 volts

$W = I^2R$

$R1 = 3a \times 3a \times 40\Omega = 360$ watts

$I = E/R \, 120v/80\Omega = 1.5$ amps

$R3 = 1.5a \times 1.5a \times 80\Omega = 180$ watts

$R2 = 1.5a \times 1.5a \times 80\Omega = 180$ watts (R2 is also 80Ω to have a total R of 20)

(a) R1 would overheat at 360 watts

FINAL EXAM Answers

9. The 12Ω and 20Ω are in parallel: $\dfrac{12 \times 20 = 240}{12 + 20 = \ 32} = 7.5Ω$

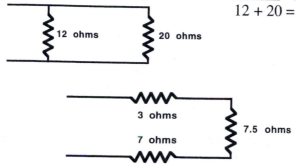

In series the resistance adds: 3Ω + 7.5Ω + 7Ω + **17.5** R total.

10. First step is to find the circuit resistance:

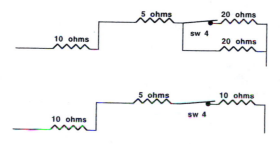

The two 20Ω resistors in parallel would reduce to a 10Ω. Total R = 25Ω.

Next step is to find I = E/R = 225v/25Ω = 9 amps current flow.

VD = I x R = 9 amps x 10Ω = **90 volts** dropped across a 10Ω resistor.

SUMMARY Answers

The first step is to find the total resistance of the circuit by combining the resistors and reducing them to the simplest form. Start at the end of the circuit by combining the 5Ω and 8Ω that are connected in parallel.

$$\frac{R1 \times R2}{R1 + R2} \qquad \frac{5\Omega \times 8\Omega = 40\Omega}{5\Omega + 8\Omega = 13\Omega} = 3.076923\Omega \text{ or } 3.077\Omega$$

Now the circuit looks like this:

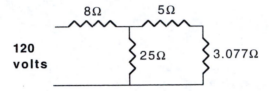

Now the 5Ω and 3.077Ω add together in series for a combined resistance of 8.077Ω which is now in parallel with the 25Ω.

$$\frac{R1 \times R2}{R1 + R2} \qquad \frac{25\Omega \times 8.077 = 201.925}{25\Omega + 8.077 = \ \ 33.077} = 6.1047\Omega$$

Now the circuit looks like this:

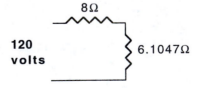

The 8Ω and the 6.1047Ω add together in series for a total circuit resistance of **14.1047** ohms.

SUMMARY Answers

Solve the total circuit current. I = E/R
120v/14.1047Ω = **8.5 amps** is the total circuit current.

The current of **8.5a** is the current flowing at **AB**.

The voltage drop at resistor **AB** is E = I x R = 8.5a x 8Ω = **68 volts** dropped. Now the voltage applied at BF is the difference of the applied 120v and the drop of 68v at AB. 120v - 68v = **52 volts** is the voltage at **BF**.

The amperage at point B is 8.5a and must divide at this point.
To find the current flow use the voltage at BF which is 52v. The resistance at BF is 25Ω and 8.077Ω.
The current flowing through BF is I = E/R = 52v/25Ω = **2.08 amps** flowing through **BF**. The current flow at BC is 52v/8.077Ω = **6.438 amps** flow at **BC**.
The voltage drop at **BC** is 6.438a x 5Ω = **32.19 vd.** The voltage at CG would be the voltage at BF 52v - 32.19v = **19.81 volts** at **CG**.

The current flow at CG is I = E/R = 19.81/8Ω = **2.476 amps** flow at **CG**.

The voltage at **DH** is also **19.81v**. The current flow at DH is I = E/R = 19.81v/5Ω = **3.96 amps** flow at **DH**.

SUMMARY Answers

Solve the following for the circuit above:

AB = __68__ volts   AB = __8.5__ amps

AE = __120__ volts   BC = __6.438__ amps

BC = __32.19__ volts   BF = __2.08__ amps

CD = __0__ volts   DH = __3.96__ amps

DH = __19.81__ volts   CG = __2.476__ amps

BF = __52__ volts   Total circuit __8.5__ amps

FG = __0__ volts

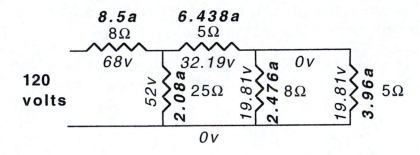

READ THE BOOKS THE ELECTRICIANS READ

WORLDWIDE LEADER IN ELECTRICAL EDUCATION

1-800-642-2633
E-mail tomhenry@code-electrical.com
ON LINE SHOPPING AT
http://www.code-electrical.com